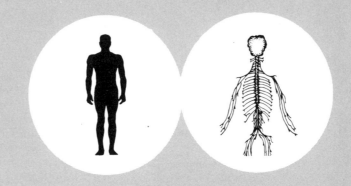

BASIC CONCEPTS
OF
ANATOMY & PHYSIOLOGY

A Programmed Study

*Programmed by the Education
and Training Department, Industrial Division,*

Bolt, Beranek and Newman, Inc.
Cambridge, Massachusetts

Under the Direction of

W. B. Dean
G. E. Farrar, Jr., M.D.
A. J. Zoldos

All of Wyeth Laboratories

J. B. LIPPINCOTT COMPANY

Philadelphia   Toronto

Cover design by Ellen Cole

Distributed in Great Britain by
Blackwell Scientific Publications
Oxford, London, and Edinburgh

ISBN-0-397-54050-7

Printed in the United States of America

9   11   10   8

Preface

Programmed textbooks are unique in that they present information to be learned in small, easily digested pieces, which in most cases require an immediate response from the reader. Errors are corrected at once, and correct answers are rewarded by the satisfaction of accomplishment. Thus the process of learning becomes nearly painless and is given an added dimension of excitement. Much accumulated experience has proved that with a well programmed text, knowledge is acquired with less effort and in less time than by standard textbooks. Programmed learning comes close to the one-student-one-teacher relationship which educators and psychologists agree is the ideal situation for learning.

Although it is a valuable learning tool by itself and can be used as a textbook, this program will also be effective for some courses if used as a supplement to a good standard textbook in anatomy and physiology in conjunction with a classroom course. Twentieth-century progress has not yet developed a substitute for a good teacher and a good textbook!

This book will be useful for a wide range of people whose occupations require a knowledge of anatomy and physiology such as nurses, laboratory technicians, medical assistants, medical secretaries, hospital corpsmen, dental technicians, commercial field representatives, college and high school students. Any others whose knowledge of these subjects needs to be refreshed and sharpened will find this a very helpful program. Originally prepared for training the personnel of a major pharmaceutical manufacturer, this thoroughly revised program assumes no prior contact with the subjects, but it leads the reader into a fairly sophisticated knowledge of the basic concepts of anatomy and physiology.

Contents

How To Use This Book

This text is programmed. Programmed instruction is a relatively new method of presenting information which has already found wide acceptance in schools, business, industry, and government. Programmed materials are materials which have been broken down into their smallest possible logical components, put back together to form a structure of interlocking logical sequences, and presented to the reader in such a way that he advances step-by-step through the subject, absorbing and accumulating information almost effortlessly.

Learning by means of programmed instruction is both remarkably effective and deceptively easy, as you will see. It also allows you to proceed entirely on your own, at your own speed, without the need for any instructor but yourself.

Information in this book is presented in short numbered paragraphs called "frames." Almost every frame will have a blank space to fill in or a multiple choice question to answer. You will know what word to fill in or what choice to select because it will be something you just learned. Write the word in the blank, or circle the correct choice (say the correct word or choice to yourself).

Terms encountered for the first time are capitalized for your convenience.

So that you can check yourself, the correct word or choice for each frame is printed next to the next frame in a shaded column on the left-hand side of the page. Use an envelope or piece of cardboard to cover this column and only expose the answer to a given frame after you have made your own answer. You will learn much more effectively if you do not get into the habit of glancing at the correct answer before you make your own.

Basic Biologic Concepts

1-1.   Human anatomy and physiology are part of <u>BIOLOGY</u> (bi – OL – o – gy), the science of all living things.

      To understand how a living substance – human, animal, plant, or whatever – is arranged and how it functions, you must study the science of ___*biology*___.

biology

1-2.   <u>ANATOMY</u> (a–NAT-o-mi) is a sub-topic of biology that describes how the body is put together so that it assumes its total <u>form</u>. How body parts are put together to assume a definite form is the study of ___*anatomy*___.

anatomy

1-3.   The human body and its parts have definite <u>form</u> and <u>structure</u>.

      Structure is how the body parts are related to each other. This relationship may refer to either the part's <u>appearance</u> or <u>position</u> in the body.

      <u>Form</u> is the three-dimensional appearance (not the outline) of the body or one of its ___*parts*___.

parts

1-4.   When you describe the body or one of its parts and you include all three dimensions, this is a description of the ___*form*___ of the body or the parts.

form	**1-5.** If you tell someone what a body part looks like and where it is located in the body, you are describing the ___*structure*___ of the part.
structure	**1-6.** Don't become confused. Form is not an outline but the three-___*dimensional*___ appearance of a box.

OUTLINE   FORM

FRONT VIEW   3-DIMENSIONAL VIEW |
| dimensional | **1-7.** A house has definite form and all of its parts should also have definite form.

When you describe the position and appearance of the windows of a house, you are referring to the ___*structure*___ of the windows. |
structure	**1-8.** The section of biology dealing with human form and structure is called ___*anatomy*___.
anatomy	**1-9.** Before the discovery of the microscope, the human body's ___*form*___ and structure, or anatomy, was examined by using the eye as the principal investigating tool.
form	**1-10.** The anatomist was able to make only a <u>gross</u> or whole examination. This scientist was able to give a gross description of the ___*whole*___ body or one of its parts.

2

whole	**1-11.** When a doctor examines a human body or any of its related parts, he is making a ___gross___ examination.
gross	**1-12.** However, there are many minute parts of the body that cannot be seen by the eye alone. The discovery of the microscope enlarged the study of anatomy. A doctor could make not only a gross examination but also a ___microscopic___ examination.
microscopic	**1-13.** Microscopic anatomy, called <u>HISTOLOGY</u> (his-TOL-o-ji) is that discipline of science which does not depend upon the unaided eye but rather upon the ___microscope___ to examine a body part.
microscope	**1-14.** When a microscopic examination of a body part is made, it is called microscopic anatomy or ___histology___.
histology	**1-15.** You might note by a gross examination that a woman was pregnant. If you wished to examine some of the affected parts in greater detail, you would use as a guide the study of ___histology___.
histology	**1-16.** If you describe the anatomy of a clock but not how it functions, you know but one-half of the story. The anatomy of the clock is important, but of equal importance is how the clock and its part work or ___function___
function	**1-17.** Function simply means how something works. The study of how the body or its parts function is called <u>PHYSIOLOGY</u> (fiz-i-OL-o-ji). A study of how the clock functions would be called clock ___physiology___.

physiology	**1-18.** The study of the function of the body can either mean how the body as a whole functions or how each individual organ functions. If you study how the stomach functions, you are studying the _physiology_ of the stomach.
physiology	**1-19.** Anatomy is the study of body form and structure and _physiology_ is the study of how the body and/or its organs function.

physiology	**1-20.** A mechanic is asked by his boss to assemble a group of parts and describe the finished product.
	
	    a. This is a study of the physiology of the machine. (move to frame 21)
	
	    b. This is a study of the anatomy of the machine. (move to frame 22)

a.	**1-21.** You said the assembling of the machine and the description of the finished product is a study of the machine's physiology. Not so! Physiology relates to how the machine and its parts function, not to their form or structure. (move to frame 22)
b.	**1-22.** You said the assembling of the parts of the machine and the description of the finished product is a study of the machine's anatomy. Right! Anatomy is a study of form (organization of the entire machine) and structure (relationship of the parts). Physiology would be a study of how the machine _functions_ (works).
	1-23. Most of us are able to recognize _normal_ body form. If, for example, the body changes shape you would not consider this _normal_.
normal	**1-24.** When you study anatomy, you are guided by establishing norms, or the normal form and _structure_ of the body.

structure	1-25. An anatomist is a scientist who establishes the guides or body _____*norms*_____ for the study of anatomy.
norms	1-26. A good auto mechanic is one who understands not only how an auto is put together but also how the engine _*functions*_ normally.
functions	1-27. To assist the mechanic, the manufacturer of the auto, like the _*physiologist*_, establishes the norms for the car.
physiologist	1-28. Manufacturers must also include a description of normal engine function. This description of the normal machine function, or engine _*physiology*_, assists in determining the efficiency of operation.
physiology	1-29. A physician, because of the norms established by the anatomist, is able to distinguish between _*normal*_ and abnormal body conditions.
normal	1-30. The study of <u>abnormal body conditions</u> is a separate medical science called <u>PATHOLOGY</u> (pa-THOL-o-ji). A pathologist is a scientist who studies abnormal _*anatomy*_ and physiology.
anatomy	1-31. The study of disease and sickness is called _*pathology*_.

pathology	1-32. To understand how a disease influences the body, the _*pathologist*_ must determine the:
	
	  1. <u>cause</u> of the disease
	  2. <u>progress</u> of the disease
	  3. <u>effect</u> of the disease on the body.

pathologist	**1-33.** An auto mechanic is like the pathologist for he must know the <u>cause</u> of the disorder, how the disorder will <u>progress</u>, and the _effect_ the abnormality will have on the operation of the auto.
effect	**1-34.** A physician is trained to treat the sick. He must, therefore, have a knowledge of the (1) _Causes_ , (2) _progress_ and (3) _effect_ of disease on the body.

1. cause
2. progress
3. effect
	1-35. A scientist who is interested in the science of <u>all</u> living things would be called a:
	
	    a. Pathologist
	    (b.) Biologist (select either a, b, or c and move to frame 36)
	    c. Anatomist

b.	**1-36.** When you visit an automobile show room and spend some time admiring a new car on display, you are in effect admiring the:
	
	    a. anatomy of the car (move to frame 38)
	    b. physiology of the car (move to frame 37)

b.	**1-37.** You said you were admiring the physiology of the car on display. Not so! Physiology refers to how the car functions or works, not to appearance. (Move to frame 38)
a.	**1-38.** You said you were admiring the anatomy of the car on display. Right. Physiology refers to function and <u>anatomy</u> refers to the (1) _form_ of the entire car and how its parts are related, or its (2) _structure_ .

1. form
2. structure
	1-39. If you examine an animal with your eyes, this study would be:
	
	    a. gross anatomy (move to frame 40)
	    b. histology (move to frame 41)

a.	**1-40.** You said the study would be gross anatomy, and with such an abbreviated description of what was being done to the animal, your answer was the only logical one. Histology could not be considered because we did not include the use of a microscope in the description of the examination. (move to frame 42)
b.	**1-41.** You said the study would be histology. But wait, was a microscope included in the description of the examination? Histology is the study of microscopic anatomy. (return to frame 40)

	1-42. While walking down a busy street, you see an attractive person walk by. What were you admiring?
	    a. her (or his) function
	    b. gross form and structure
	    c. Histologic form and structure
	
	(select either a, b, or c and move to frame 43)

b.	**1-43.** This is a tough one to defend, but let's allow logic to guide us. First, I am sure you did not want (intend) to look at her (or him) under a microscope, so histology is out. Second, you probably were not interested (or were you) in how her (or his) organs worked, so function is out. (no response required - move to frame 44)
	1-44. You were undoubtedly impressed with the form of the <u>entire</u> body. There is also a good possibility that you were impressed by the structure, or how the parts were related to one another. Because your examination was done by using only the unaided eye, this was a study of the person's ____gross____ anatomy.
gross	**1-45.** Before you attempt to repair a machine, it is best to know how it functions normally. The study of machine function is called machine _physiology_.
physiology	**1-46.** If you understand the normal anatomy and physiology of a machine and its parts, you can predict how it will _function_ if a part is broken.

7

function	**1-47.** Physicians examine a sick patient to locate any abnormalities of either the anatomy or physiology. If the doctor discovers disease, he must determine the cause, progress, and _effect_ of the disease on the body.
effect	**1-48.** The study of the cause, progress, and effect of the disease on the body is called _pathology_.
pathology	**1-49.** You now have finished the first section of this unit. Before you go on you may want to take a short break.
	1-50. All natural sciences are interrelated -- you should expect biology, chemistry, and physics to overlap in essential areas. An understanding of chemistry will help you better understand human physiology. (no response required - move to frame 51)
	1-51. General chemistry has many sub-topics among which is the chemistry of life, or <u>BIOCHEMISTRY</u> (bi-o-KEM-is-tree). Biology is the study of all living things and <u>BIO</u>chemistry is the chemistry of _living_ things, or life chemistry.
living	**1-52.** You are constantly being influenced by chemical compounds and elements. For example, if you did not have ample water (H_2O) you would die. It is, therefore, important that you understand how _biochemistry_, the chemistry of living things, is organized.
biochemistry	**1-53.** If you were asked to name some chemicals, I am sure you could easily do so. You might say sodium, gold, oxygen and any number of common chemicals. These units of chemistry are called <u>chemical elements.</u> Hydrogen is such a unit or _chemical_ element.

chemical	1-54. Some philosophers have stated that man is distinguished from all other living creatures by his use of symbols. This may be true when one considers we have a word to describe our every action. Chemistry, too, has symbols for each element, and together these symbols form a chemical language. For example, Fe is the symbol for the _chemical_ _element_ iron.
chemical element	1-55. There are 103 (as of 1966) known chemical elements. Each element is a single substance and has a separate chemical _symbol_ indicating what element it is.
symbol	1-56. A chemist must be able to use these symbols, as you do letters of the alphabet. You are not expected to master the chemist's language, but you should remember that the simplest chemical substance is an element and each element is represented by a _symbol_.
symbol	1-57. Fe – P – Na These symbols represent the simplest form of the chemical _elements_, iron, phosphorus and sodium.
elements	1-58. Scientists wondered why some elements <u>combined</u> and others would <u>not</u>. The research of many scientists finally resulted in the discovery of the building blocks of elements -- the <u>ATOM</u> (at-om). Elements are, therefore, made up of one or more of these building blocks called _atoms_.
atoms	1-59. When the chemical element gold is broken down into its smallest part that still physically resembles the element gold and reacts as the gold element normally does with other elements, this particle is an _atom_ of gold.
atom	1-60. There are 103 known chemical elements so there must also be 103 different building blocks or _atoms_, one for each of the chemical elements.

9

atoms	1-61. You would, if mining metals, soon be able to recognize the difference between gold and lead. It is not hard to conceive, therefore, that the difference between these elements (would/ would not) (choose one) _____would_____ be because of the fundamental difference in their building blocks, the atoms.
would	1-62. To summarize, you have discovered that: a. the smallest part of an element that can exist is an (1) _____atom_____. b. there are as many different kinds of atoms as there are (2) _____elements_____.
1. atom 2. elements	1-63. The word HAT can be broken down into the simple symbols (letters) H-A-T. The chemical word H_2O (water) can be broken down into three symbols, H-O-H, to represent two _____atoms_____ of hydrogen and one atom of oxygen.
atoms	1-64. You may have seen a druggist compounding a medicine. He mixes <u>two or more</u> drugs together to _____compound_____ the medicine.
compound	1-65. In its simplest terms a <u>chemical compound</u> is a combination of two or more different elements. H + O + H = HOH or H_2O atom atom atom = compound When H + O + H are combined, the _____compound_____ water is formed.
compound	1-66. Indicate which are elements and which are compounds (e for element, c for compound). a. Fe_____e_____ c. P_2O_5_____c_____ b. MnO_2_____c_____ d. Na_____e_____
a. e b. c c. c d. e	1-67. If you combine an atom of sodium (Na), plus an atom of chlorine (Cl), you will produce the chemical _____compound_____ table salt (NaCl). Each combined form (NaCl) is also called a <u>MOLECULE</u> (MOL-e-kul) of salt.

compound	1-68. The chemical compound water, HOH, is composed of three atoms (H-O-H). HOH is a molecule of water. NaC1 is the compound table salt -- it is also a ___molecule___ of salt.
molecule	1-69. The smallest part of an element is an atom. Because compounds are made by combining the atoms of elements, the smallest part of a compound would be a ___molecule___
molecule	1-70. This symbol (Fe) represents an (1)___atom___ of the element iron. This formula (MnO$_2$) tells you that the elements Mn and O$_2$ have combined to form the compound MnO$_2$, the smallest part of which is a (2)___molecule___.
1. atom 2. molecule	1-71. During the early 1900's, it was thought that the atom was the smallest particle that could exist. You know that this is no longer considered to be true for during World War II the development of the ___atom___ bomb was accomplished when the atom itself was split.
atom	1-72. But let's define what is meant by a PARTICLE (PAR-ti-cle) that is smaller than an atom. An atom is composed of three particles, the electron, the proton and neutron. (No response required - move to next frame).
	1-73. The hydrogen atom is a simple atom having only two of these small parts or particles, an ___electron___ and a proton.

11

electron	**1-74.** The compound water (HOH) can be broken down into the atoms (H-O-H). These atoms are composed of small parts or _particles_.
particles	**1-75.** The atom is the smallest part of an element that contains all the properties (how the element reacts and what it looks like) of the element. The particles of the atom are not small bits of the _element_ but parts of the atom itself.
element	**1-76.** This atom has three different particles, neutrons, _electrons_ and positive _protons_. HELIUM ATOM 1. ELECTRONS NUCLEUS (−) (+) ○ (+) (−) 2. PROTON 3. NEUTRON (NU-tron)
electrons protons	**1-77.** All atoms (except hydrogen) are composed of three basic particles, the electron, the proton and the particle with a neutral charge, the _neutron_.
neutron	**1-78.** Atoms have _electrons_ rotating about the positive _protons_ and neutral neutrons.
electrons protons	**1-79.** The atomic particles, except the neutron carry an electric charge; the proton, a positive (+) charge, and the electron, a negative (−) charge. For example, a lithium atom has three protons or three _+_ charges and three electrons or three _−_ charges.

positive or (+) negative or (-)	1-80. A beryllium atom has four _proto___ or four (+) charges and four electrons or ___four___ negative charges.
protons (+) four	1-81. Atoms are physically different because each atom's nucleus contains a different number of neutrons and ___protons___.
protons	1-82. While the electron that is whizzing about the nucleus does not weigh as much as a proton (an electron weighs 1/1847 of one proton), it does influence how an atom will mix with its neighbors. So the chemical activity of an atom is dependent upon the rotating ___electron___.
electrons	1-83. Elements differ in physical characteristics and also differ in how they react chemically. Chemical reaction is controlled by the ___electrons'___ action, and the physical characteristics are controlled by the ___protons___ and neutrons in the nucleus.
electrons' protons	1-84. A lithium atom that contains three (+) positive charges (protons) and three (-) negative charges (electrons) is electrically neutral (carries no charge) because the (+) charges equally balance the ___—___ charges.
(-) or negative	1-85. If an atom has the same number of (-) and (+) charges, it (will/will not) ___will not___ have an electric charge.

will not	1-86. Atoms with opposite charges are attracted to each other.
	(-) atom ___attracts___ (+) atom
	
	    Atoms with no charge or the same charge are not attracted to each other.
	
	    Negatively-charged atoms (may/may not) ___may___ combine with positive-charged atoms.

13

attracts may	1-87. When a neutral atom loses a (–) charge it has one more (+) charge and it then becomes a ___+___ charged atom.
(+) positively	1-88. The element oxygen has eight (+) charges and eight (–) charges and is, therefore, neutral. If, however, the oxygen element gains two electrons it will have a total of ten negative charges and will become ___–___ charged.
(–) or negatively	1-89 A charged atom is called an ION (i–on). An atom carrying an excess of electrons (more electrons than it has protons in the nucleus) is an example of an ___ion___.
ion	1-90. It is important to remember that an atom can become charged by gaining or losing only its electrons. The protons are held within the ___nucleus___ and are not able, under normal conditions, to leave the atom.
nucleus	1-91. Count the number of electrons and protons. This atom is an ___ion___ or positively charged atom.
ion	1-92. This atom has three electrons and four protons. It is, therefore, an ion with a ___+___ charge.
(+) positive	1-93. You can define an ion as an atom that has gained or lost one or more of its ___electrons___.

electrons	1-94. A negative ion may attract a _____+_____ ion and form a chemical compound.
positive	1-95. Chemists have termed ions as either ANIONS (AN-i-on-s) (-) charged or as CATIONS (KAT-i-on-s) (+) charged. An atom that has lost one or more electrons is a CATION because it now carries a _____+_____ charge.
(+)	1-96. Atoms that gain one or more electrons are called _anions_ because they now have a (-) charge.
anions	1-97. An ANION is attracted toward a _cation_, a (+) charged atom.
cation	1-98. If you place an electrically (-) pole in a solution, it will attract the (+) _cations_.
cations	1-99. Negative ions or (1) _anions_ will be attracted to the (+) pole. Cations will be attracted to the (2) __/___ pole.
1. anions 2. (-)	1-100 This is a _cation_ because it has a charge of one more proton than the total number of electrons

15

cation	1-101. An atom with six electrons and four protons would be an anion with a charge of (-2), because it has ___2___ more electrons than the total number of protons.
2 (two)	1-102. You must master this chemical language now for later in the program when we refer to the fluid environment of the human body we will discuss how these ions combine to form chemical ___compounds___.
compounds	1-103. If you could return to the "Dawn of Life", you would find _all_ forms of life surrounded by an environment of sea water. As the surface of the earth changed, the fluid environment of living things also changed. Some living things remained in the fluid environment while others enclosed this fluid within their bodies. (move to the next frame)
	1-104. Today, human blood serum closely resembles the sea water of the early periods of the earth's history. Man no longer lives in a fluid environment but carries this environment ___within___ his body.
within	1-105. This fluid environment within the body contains ions. When ions are placed in water, an electrical current can be passed through the solution. Such a solution is called an ELECTROLYTE (e-LEC-tro-lite) solution and is so because it contains ___ions___.
ions	1-106. Dry table salt will not easily allow electricity to pass through it (conduct electricity). However, when (NaC1) table salt is dissolved in water, it dissociates (breaks down) into (Na +) and (C1-) ions. The solution can now be called an electro ___lyte___.
lyte	1-107. Water is a poor electrolyte. However, if a chemical is dissolved in the water and it breaks down (dissociates) into ___ions___, the water becomes a good conductor.

ions	1-108. Compounds that break down into ions in solution make the solution a good conductor of electricity or an _electrolyte_ solution.
electrolyte	1-109. The blood serum, <u>one</u> of the fluid environments of the body, permits compounds to dissociate into ions and is, consequently, an _electrolyte_ solution.
electrolyte	1-110. The ions that dissociate in your blood serum are called <u>ELECTROLYTES</u> (e-LEC-tro-lites). When table salt is put into a solution and breaks down into (Na +) and (C1-) ions, these should be called the _electrolytes_ of the solution.
electrolytes	1-111. When your body contains too few or too many electrolytes, sickness or death can follow. If you take a salt-water fish out of its environment and place it in fresh water, it will generally _die_.
die	1-112. The fish was originally in the environment of an electrolytic solution with a (higher/lower) _higher_ concentration of salt ions.
higher	1-113. When the _electrolytic_ balance of the fish's environment was changed, sickness and death followed.

electrolytic	1-114. It is extremely important that the <u>concentration</u> of <u>electrolytes</u> in solution and the <u>ratio</u> of <u>solution</u> to the available electrolytes be maintained at a normal level.
	
	        If this <u>balance</u> is upset it becomes difficult to maintain good health.
	
	        (move to next frame)

17

1-115. You have been through a rather concentrated section of basic chemistry. A break is in order.

1-116. The smallest part of an element containing all the properties of the element is:

  a. an atom (move to frame 117)

  b. a molecule (move to frame 118)

---

a.

1-117. You said an atom was the smallest part of an element containing all the properties of the element. Right!

An element is made up of one type of atom only.

(move to frame 119)

---

b.

1-118. You said a molecule was the smallest part of an element containing all the properties of the element. Not so!

A molecule is the smallest part of a compound.

(move to frame 117)

1-119. Elements are pure substances that cannot, by normal chemical means, be broken down into parts smaller than an ___atom___ .

---

atom

1-120. H_2SO_4

The above figure is an example of a:
  a. symbol
  b. element
  c. compound
(select either a, b, or c and move to the next frame)

---

c.

1-121. Symbols are the letters chemists use to denote elements – sodium (Na) – iron (Fe).

A compound or molecule is made by combining two or more ___elements___ .

18

elements	1-122. The number of electrons that revolve about the nucleus of an atom influence how an element will ___*react*___ with other chemical elements.
react	1-123. Protons and neutrons that are confined within the nucleus of of an atom determine the ___*physical*___ appearance of the atom.

physical	1-124. Atoms with eight electrons and six protons are called:
	
	      a. cations (move to frame 126)
	      b. anions (move to frame 125)

b.	1-125. You said atoms with eight electrons and six protons are called anions. Right!
	
	    8 electrons = −8
	    6 protons = <u> +6 </u>
	                 −2 charge
	This atom will go to the (+) charged pole. (move to frame 127)

a.	1-126. You said atoms with eight electrons and six protons are called cations. Not so! (return to frame 124)
	1-127. When a compound is dissociated in a solution it forms ions called ___*electrolytes*___
electrolytes	1-128. If common table salt (NaCl) is put into solution, it dissociates into sodium (Na+) or ___*cations*___ (+ ions) and chlorine (Cl−) or ___*anions*___.

cations anions	1-129. An electrolytic solution such as blood serum is one which contains electrolytes and will allow ___electricity___ to pass through it.
electricity	1-130. If the electrolytic solution of your body becomes unbalanced (too many or too few electrolytes in solution), it is difficult to maintain good ___health___ .
health	1-131. Chemical compounds that are NOT produced by living things are called INORGANIC (in-or-GAN-ik). These compounds constitute the ASH of the body. The ash of the body contains INORGANIC chemical ___compounds___ .
compounds	1-132. Chemical compounds that ARE made by living substances are called ORGANIC (or-GAN-ik). Ninety-six per cent of your body is composed of ORGANIC compounds; the remaining four per cent are not made by your body and are, therefore, ash or ___inorganic___ compounds.
inorganic	1-133. Plants manufacture simple sugars. These compounds, because they are produced by living things, are ___organic___ compounds.
organic	1-134. Compounds of calcium, sodium, potassium and chlorine that are not manufactured by living things are ___inorganic___ compounds.
inorganic	1-135. The three main organic compounds of the body are CARBOHYDRATES (kar-bo-HI-drates) (sugars and starches), LIPIDS (LI-pids) (fats) and PROTEINS (PRO-te-ins). The protein of meat is an ___organic___ compound.

organic	1-136. Carbohydrates, lipids, and proteins, are all organic chemicals because they are made by _living_ things.
living	1-137. The three main organic chemical compounds used by the body as the major sources of food are _carbohydrate_ _lipids_, and _proteins_.
lipids (fats); carbohydrates (sugars); proteins (any order)	1-138. You may accurately describe man as a blob of ""living stuff". A more scientific term for "living stuff" is PROTOPLASM (PRO-toe-plazm). This "living stuff" or _protoplasm_ is composed of both organic and inorganic compounds.
protoplasm	1-139 All living things, plants and animals, are composed of this living material called _protoplasm_.
protoplasm	1-140. Blood cells, skin and bone (are/are not) _are_ all forms of protoplasm.
are	1-141. Protoplasm is a colorless viscous material that is in a semi-liquid state because of the particles suspended in it. Try to imagine protoplasm as being like raspberry jam. The raspberry seeds would be the particles that are _suspended_ in the protoplasm.
suspended	1-142. Inorganic compounds containing sodium, calcium, magnesium, potassium and _inorganic_ compounds of lipids, proteins and carbohydrates are found in protoplasm.

21

organic	1-143. So far, we have talked about living things, but what does it really mean "to be alive?" The basic characteristics of a living substance are things you do every day, such as moving, eating and responding to stimuli. (no response required; move to frame 144)
living	1-144. Without being prompted or prodded, protoplasm can move from one area to another. This property of protoplasmic movement is called MOTILITY (mo-TIL-i-ti). MOTILITY is one of the life functions that is characteristic of all _____living_____ things.
motility	1-145. A piece of equipment may be mobile or easily moved. Such equipment depends upon an outside force for its movement. Protoplasmic motion or _____motility_____ can take place without the assistance of an outside force.
motility	1-146. Some protoplasm moves by a flowing action similar to jam moving down the side of the jar. At other times protoplasm depends upon small hair-like oars for its _____motility_____ .
food	1-147. Have you noticed that many times you are not hungry until you look at the clock? The clock is the STIMULUS (STIM-u-lus) that caused you to act and in this case to think of _____food_____ or eating.
stimulus	1-148. The process of responding to a stimulus is termed IRRITABILITY (ir-i-ta-BIL-i-ti). Protoplasm exhibits this function of irritability or responding to a _____stimulus_____ .
	1-149. Environmental changes are an example of a stimulus that can cause an _____irritability_____ reaction in a living thing.

22

irritability	1-150. When protoplasm responds to a stimulus, the protoplasmic reaction is called ___irritability___.
irritability	1-151. Protoplasm, when stimulated, sets up a wave motion that is similar to a stone's being dropped into a pool of water. This wave motion carries the ___stimulus___ to every section of the protoplasm.
stimulus	1-152. There are many stimuli that may affect protoplasm and cause it to respond. Heat, electricity, or mechanical stimuli may induce the protoplasmic response called ___irritability___
irritability	1-153. Protoplasm is truly dynamic -- it never rests. Changes within the protoplasm are occurring continually. Some of the protoplasmic material is being built up and other portions broken down. Protoplasm is in a constant state of ___change___.
change	1-154 This process of construction and destruction within protoplasm and the removal of waste is called METABOLISM (me-TAB-o-lis-m). Some of the metabolic activity is ___constructive___ while another phase is devoted to the destruction of the protoplasm
constructive	1-155. When protoplasm is in the constructive phase, it is termed ANABOLISM (a-NAB-o-lizm). New protoplasm is made during the anabolic phase of ___metabolism___.
metabolism	1-156. Metabolism has a destructive phase. When protoplasm is being destroyed, it is called CATABOLISM (ca-TAB-o-lizm). Protoplasm is being broken down during the: a. anabolic phase. (move to frame 158) b. catabolic phase. (move to frame 157)

b.	**1-157.** You said catabolism is the phase of metabolism in which protoplasm is being broken down. Right! It is during catabolism that protoplasm is broken down to release energy for body functions such as irritability. (move to frame 159)
a.	**1-158.** You said anabolism is the phase of metabolism in which protoplasm is being broken down. Not so! (move to frame 159)
	1-159. During anabolism, protoplasm is being built up to balance the action of _catabolism_, or destruction of protoplasm.

catabolism	**1-160.** The greatest protoplasmic growth would take place during:
	
	    a. catabolism (move to frame 161)
	
	    b. anabolism (move to frame 162)

a	**1-161.** You said that the greatest growth would take place during catabolism. This is not so! (move to frame 162)
b	**1-162.** You said that the greatest growth would take place during anabolism. Right! During the process of anabolism, protoplasm is used and new protoplasm made. Remember, anabolism means to build up -- catabolism, to tear down.
	1-163. It is not difficult to imagine what would happen to the body if the process of catabolism were more active than the process of anabolism. The body (would/would not) _would_ be in a declining or unhealthy state.

would	1-164. As protoplasm is broken down during catabolism to produce energy, the waste formed must be eliminated. This process of elimination of waste, or EXCRETION (eks-KRE-shun), is another metabolic activity of all _____ things.
living	1-165. Waste is sifted out and eliminated from the body by the process called _excretion_.
excretion	1-166. The activities that all living substances must exhibit are: a. spontaneous movement, or 1. _motility_ b. response to a stimulus, or 2. _irritability_
1. motility 2. irritability	1-167. c. metabolism, or the 1. _anabolism_ and 2. _catabolism_ of protoplasmic material, and the selection and elimination of waste or 3. _excretion_.
1. anabolism 2. catabolism 3. excretion	1-168. There is a fourth life function -- that of perpetuating protoplasm by growth and division to make more protoplasm. This function of producing more of one's kind is called _reproduction_.
reproduction	1-169. Compounds that comprise four per cent of the body and are not made by the body are _inorganic_ chemical compounds, or the ash of the body.
inorganic	1-170. Carbohydrates, lipids and proteins are all examples of _organic_ chemical compounds.

organic	1-171. All living things are composed of the "living stuff" called _protoplasm_.
protoplasm	1-172. Protoplasm is a combination of water, inorganic and organic compounds. You could describe its physical appearance as being a colorless, viscous material in a _semi-liquid_ state.
semi-liquid	1-173. The total sum of the anabolic and catabolic activity plus excretory action is called _metabolism_.
metabolism	1-174. You can classify a substance as being alive if it performs the following four life functions: 1. motility 2. irritability 3. metabolism 4. _reproduction_
reproduction	1-175. The biologic language that you have studied in unit one will be used in every unit of this course. For example, unit two will contain information about cells and tissues, which are specialized forms of protoplasm. Cells have a definite anatomy, and their physiology is dependent upon the actions of ions, organic, and inorganic compounds. Like all living substances, cells and tissues must carry out the four life functions.
	You should take a break of an hour or more before beginning Chapter II – The Human Cell.

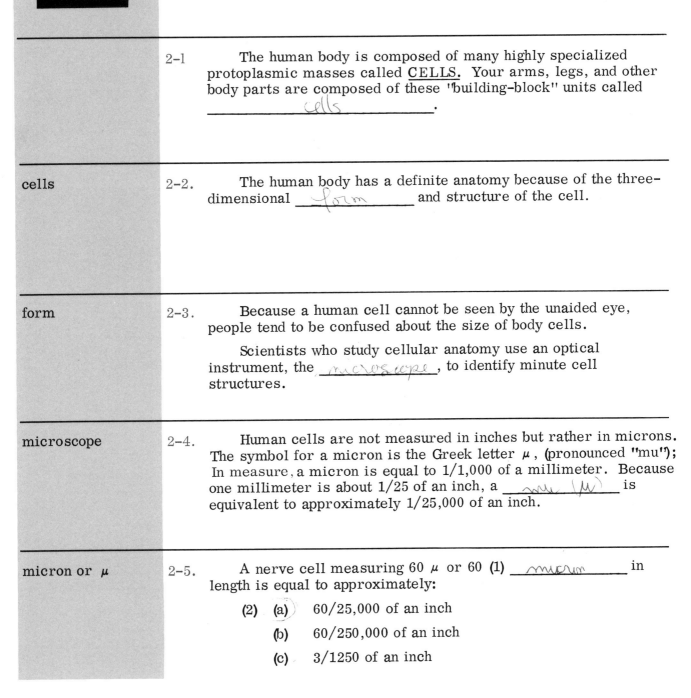

CHAPTER 2

The Human Cell

2-1   The human body is composed of many highly specialized protoplasmic masses called <u>CELLS.</u> Your arms, legs, and other body parts are composed of these "building-block" units called _____cells_____.

---

cells

2-2.   The human body has a definite anatomy because of the three-dimensional ___form___ and structure of the cell.

---

form

2-3.   Because a human cell cannot be seen by the unaided eye, people tend to be confused about the size of body cells.

    Scientists who study cellular anatomy use an optical instrument, the ___microscope___, to identify minute cell structures.

---

microscope

2-4.   Human cells are not measured in inches but rather in microns. The symbol for a micron is the Greek letter μ, (pronounced "mu"); In measure, a micron is equal to 1/1,000 of a millimeter. Because one millimeter is about 1/25 of an inch, a ___micron (μ)___ is equivalent to approximately 1/25,000 of an inch.

---

micron or μ

2-5.   A nerve cell measuring 60 μ or 60 (1) ___micron___ in length is equal to approximately:

    (2)   (a)   60/25,000 of an inch

           (b)   60/250,000 of an inch

           (c)   3/1250 of an inch

	2-6. The average size of a human body cell is 20 microns which can also be written using the micron symbol as 20 ___*μ*___ .
	2-7. It is difficult to conceive what a micron really means, or to visualize accurately how small many microscopic organisms are. However, most single-celled bacilli (a form of bacteria) measure about one micron. This means that it would take more than 1,500 of these bacilli laid side by side to span 1/16 of an inch! (No response required – move to frame 8)
	2-8. You have probably noted that a single isolated soap bubble is spherical. When a single soap bubble is "sandwiched" in among many soap bubbles, it is no longer spherical but compressed into an ___irregular___ shape. SPHERICAL IRREGULAR
	2-9. A single isolated body cell can be compared to an isolated soap bubble because it has a ___spherical___ ___form___ and is three-dimensional.
	2-10. The form of an isolated cell may not <u>always</u> be spherical. For example, the forms of those specialized cells which have ___specialized___ body functions to perform are <u>not</u> spherical. NERVE CELL MUSCLE CELLS

specialized	2-11.	Because of printing limitations, the cells illustrated in this program are drawn in two dimensions. You must remember however, that your body cells have form and structure and are all ___3 - D___ .

ANATOMY OF A BODY CELL

three-dimensional	2-12.	

CELL MEMBRANE

CELL

Every body cell, whether it has a spherical or an elongated form, is surrounded by a protective cover, the cell _membrane_ .

---

membrane	2-13.	The cell membrane controls the transfer of body substances <u>into</u> or <u>out of</u> a cell. For example, a cell membrane allows nutrients to pass into a cell and waste to pass out of the cell. Those substances which normally <u>stay</u> within the cell are kept inside because of ___cell___ ___membrane___ action.

---

cell membrane	2-14.	A membrane that permits only selected substances to pass through it is called a <u>SEMI-PERMEABLE</u> (PUR - me - a - bl) membrane. All body cells have semi-_permeable_ membranes.

---

permeable	2-15.	Nerve cells and muscle cells, like other body cells, have a _semi-permeable_ cell membrane.

---

semi-permeable	2-16.	CELL

MEMBRANE

CYTOPLASM
(SY - to - plazm)

Within the cell membrane is the "machine-shop" of the cell, called the _cytoplasm_ .

cytoplasm	2-17. The cell functions of motility and irritability are carried out by this cell "machine-shop", the ___cytoplasm___.
cytoplasm	2-18. You know that the cell is a mass of protoplasm, or living substances. The individual parts of a cell are composed of the same material as the whole, but because these parts perform <u>specialized</u> functions, they are said to be a specialized form of ___protoplasm___
protoplasm	2-19. The cell cytoplasm is an example of ___specialized___ protoplasm.
specialized	2-20. Within the cytoplasm itself are living substances, the <u>ORGANOIDS</u> (OR - gan - oids), and also <u>non-living</u> substances, the <u>INCLUSIONS</u> (in - KLOO - zhuns). The organoids are another example of a specialized form of protoplasm and are therefore ___living___ (living/non-living).
living	2-21. The <u>non-living</u> substances within the cytoplasm are called inclusions, and the living substances are called ___organoids___.
organoids	2-22. Cytoplasm contains organoids and several non-living substances called ___inclusions___.
inclusions	2-23. CYTOPLASM 1. ___organoids___ 2. ___inclusions___ LIVING SUBSTANCE NON-LIVING SUBSTANCE Fill in the blanks.

(1) organoid (2) inclusion	2-24. Organoids within the cytoplasm have different functions. For example, the function of one kind of organoid, the <u>MITOCHONDRIA</u> (mit-o-KON-dri-ah), is to convert the <u>stored</u> chemical <u>energy</u> of the cell into a more usable form of <u> energy </u>.
energy	2-25. Whenever it is necessary to take the stored energy of the cell out of storage and convert it to use, this job is done by the <u> mitochondria </u>
mitochondria	2-26. Within the cell's energy-converters, or mitochondria, are tiny particles called <u>GRANULES</u> (GRAN-ules). These tiny particles or <u> granules </u> of mitochondria are responsible for liberating energy.
granules	2-27. The mitochondria are organoids and are therefore <u> living </u> (living/non-living) substances.
living	2-28. The inclusions, or non-living substances, perform still another set of functions. Inclusion granules can take the form of a substance used to bind oxygen for transport through the cell, or to store food. Carbohydrates, fats, and proteins can be stored in these inclusion <u> granules </u>.
granules	2-29. The inclusion granules may also be parts of dead cells, debris, or simply dust. The waste material of a cell is located in the <u> inclusion </u> granules.
inclusion	2-30. To sum up: cellular waste, oxygen and food storage are all contained in the non-living material, or <u> inclusion </u> granules.

31

inclusion	2-31.	Both inclusions and organoids are contained in the cytoplasm; the organoids that control the conversion of the cell's energy are the _mitochondrion_.
mitochondria	2-32.	Before you go on, review what you have learned so far. An isolated body cell is roughly _spherical_ in form.
spherical	2-33.	The size of an average human body cell is approximately 20 _microns_, or 20/25,000 of an inch.
microns	2-34.	Every human body cell is surrounded by a cell _membrane_.
membrane	2-35.	The "machine-shop" within the cell membrane is a type of specialized protoplasm called _cytoplasm_.
cytoplasm	2-36.	The cytoplasm carries out the life functions of _motility_ and _irritability_.
motility irritability (either order)	2-37.	The cytoplasm contains living substances called _organoids_ and non-living substances called _inclusions_.

32

organoids inclusions	2-38. Among the organoids are the cell's energy-converters, the _mitochondria_ granules.
mitochondria	2-39. Food, oxygen, and <u>waste</u> are stored in _inclusion_ granules.

inclusion	2-40. Granules are:
	    a. living
	    b. dead
	    (c.) can be either living or dead

c.	2-41. Near the center of the cell is an egg-shaped body, the NUCLEUS (NU-cle-us). The word nucleus comes from a Latin word meaning the kernel of a nut. In the same way that the kernel of a nut is within the nut, the _nucleus_ of a cell is also located inside the cell proper. nucleus (Fill In) CELL
nucleus	BODY CELL 2-42. The arrow in this drawing points to the _nucleus_ of the cell. BODY CELL

33

nucleus	**2-43.** A <u>BINUCLEATE</u> (bi-NU-kle-ate) cell is one with two nuclei (nuclei is the plural of nucleus). A <u>MULTINUCLEATED</u> (multi-NU-kle-ated) cell is a cell with many _nuclei_ .
nuclei	**2-44.** While most cells have only one nucleus, there are cells with two nuclei, binucleate cells, and cells with many _nuclei_ , or multinucleated cells.
nuclei	**2-45.** The nucleus acts as a "control center" for the vital activities of the cell such as growth, reproduction, and the development of new cells. If the body requires new cells to repair damage, the _nucleus_ will initiate this action.
nucleus	**2-46.** The living substance within the nucleus, called NUCLEOPLASM (NU-kle-o-plazm), is somewhat <u>more dense</u> than the remaining protoplasm of the cell. Nuclear sap is one of the elements of the ~~nucleoplasm~~. nucleoplasm
nucleoplasm	**2-47.** A pliable membrane, the <u>NUCLEAR MEMBRANE</u>, surrounds the _nucleoplasm_ or living nuclear substance.
nucleoplasm	**2-48.** Of the materials which compose the nucleoplasm, the most important ones to remember are the <u>NUCLEOPROTEINS</u> (NU-kle-o-pro-te-ins). Nucleoproteins are compounds of _proteins_ and <u>nucleic</u> (nu-KLE-ic) acids.
proteins	**2-49.** Proteins and nucleic acids combine to form the _nucleoproteins_

nucleoproteins	2-50. Here is a chart to help you picture the composition of the nucleoproteins: NUCLEOPROTEINS PROTEINS NUCLEIC ACIDS 1. RNA 2. DNA
	2-51. The nucleic acids are divided into two types, called <u>RNA</u> and <u>DNA</u>. The full names of these acids are <u>ribonucleic acid</u> and <u>deoxyribonucleic acid</u>, but for the sake of simplicity they are usually referred to by their abbreviations, ____*RNA*____ and ____*DNA*____.
RNA, DNA (either order)	2-52. RNA and DNA are both classified as __*nucleic*__ acids.
nucleic	2-53. Research done by Avery in 1944 indicates that DNA plays a role in the transmission of heritable variations. The characteristics you inherit from your parents appear to be influenced by the nucleic acid ____*DNA*____.
DNA	2-54. RNA functions primarily in the cytoplasm of the cell as a pattern for the formation of specific cellular proteins. There is evidence, however, that this ____*protein*____ formation function, though carried out by RNA, may be controlled by DNA.
protein	2-55. DNA is found in all nuclei, and it is confined to the dark heavy thread-like substances of the nucleus called CHROMOSOMES (KRO-mo-soms). Human body cells have <u>46</u> of these thread-like ____*chromosomes*____.

chromosomes	2-56. **NUCLEUS** 1. *nucleoplasm* 2. (46) *chromosomes*
(1) nucleoplasm (2) chromosomes	2-57. The chromosomes within the nucleoplasm carry the cell's <u>HEREDITY</u> (he-RED-i-ty) -- characteristics such as size, shape, and number of nuclei in the cell. When a cell reproduces -- by dividing in two -- the ___*chromosomes*___ which carry the hereditary characteristics, are split.
chromosomes	2-58. Let's review what you know about the nucleus. The protoplasm of the nucleus is called the *nucleoplasm* and is surrounded by a ___*nuclear*___ ___*membrane*___.
nucleoplasm nuclear membrane	2-59. The most important materials of the nucleoplasm are the *nucleoproteins* composed of nucleic acids and proteins. 1. NUCLEO*plasm* 2. *nucleoproteins*
nucleoproteins (1) plasm (2) nucleoproteins	2-60. The nucleic acid that influences heritable variations is ___*DNA*___. The nucleic acid that functions in the cytoplasm is ___*RNA*___.
DNA RNA	2-61. The cell's hereditary characteristics are carried by thread-like substances within the nucleus, called ___*chromosomes*___.

chromosomes	**2-62.** Cell theory states that:
	
	    1. All _living_ things are composed of cells.
	    2. All cells are a mass of _protoplasm_.
	    3. All cells exhibit the same four life functions.

living protoplasm	**2-63.** The life function of reproduction indicates that all new cells come only from previously existing _cell_.
cells	**2-64.** When the body parts are repaired, or the body itself grows, the cell must exhibit the life function of _reproduction_.
reproduction	**2-65.** The repair of skin after a superficial injury and the growth of hair and fingernails are all examples of cell _reproduction_.
reproduction	**2-66.** Most cells reproduce when they are mature. An exception is the red blood cell, which loses its nucleus in maturity, and must, therefore, reproduce while it is <u>immature</u>. But <u>sometime</u> during their life histories, _all_ body cells are able to reproduce.
all	**2-67.** Each cell reproduces by dividing roughly in half to form two new cells. The normal process by which cells _divide_ to make new cells is called <u>MITOSIS</u> (my-TOE-sis).
divide	**2-68.** Mitosis proceeds by a series of definite, orderly phases. At the end of the series the one old cell has been divided into two entirely new cells, and the process of _mitosis_ is completed.

37

mitosis	2-69. When cells are not exactly engaged in reproduction, they are said to be <u>resting</u>. They can be compared to the engine of your car while <u>it is idling</u> at a stoplight: it is performing all the functions of running except moving you and the car forward. When cells are resting, they are performing all the normal functions of life <u>except</u> passing through the phases of _____*mitosis*_____.
mitosis	2-70. When cells are not reproducing, but are otherwise carrying out all the normal life functions, they are said to be in the _____*resting*_____ phase.
resting	2-71. Mitosis actually begins with the <u>PROPHASE</u> (PRO-phase). The 46 chromosomes within the nucleus become <u>thick</u> and <u>more</u> <u>conspicuous</u> as the cell enters this first phase, the _____*prophase*_____.
prophase	2-72. Cell A is in the resting phase. The thickening of the chromosomes in cell B indicates that it has entered the _____*prophase*_____.
prophase	2-73. As the prophase continues, the individual chromosomes split lengthwise into <u>pairs</u>, and the nuclear membrane disappears. The 46 original chromosomes are now _____*92*_____ in number, arranged in 46 pairs.

38

2-74.

CELL A   CELL B

PROPHASE

Cell A has just entered the prophase. In cell B the chromosomes have ___split___ , and the __nuclear membrane__ has disappeared.

---

split

nuclear
membrane

2-75. After the prophase is completed, the <u>second</u> stage of mitosis, the <u>METAPHASE</u> (MET-a-phase), begins. During the metaphase, the chromosomes become attached to a delicate system of fibre-like substances which have arisen between the opposite poles of the cell. The chromosomes are suspended midway between these 2 ___poles___ of the cell.

---

poles

2-76.

METAPHASE

This drawing shows the chromosomes attached to the fibre-like substances extending from the poles of the cell during the ___metaphase___.

---

metaphase

2-77. When the nuclear membrane disappears and the 46 chromosomes split into pairs, this stage is termed the late ___prophase___

---

prophase

2-78. The stage after the prophase, when the chromosomes are attached to the fibre-like substances extending from the poles of the cell, is called the ___metaphase___.

metaphase	2-79. The <u>third</u> phase of mitosis is the <u>ANAPHASE</u> (AN-a-phase). During the anaphase, the halves of the split chromosomes finally separate completely and move toward the opposite poles of the cell, one half to one pole and the other half to the other, until there is a full set of _____46_____ (number) chromosomes at each pole.
46	2-80. ANAPHASE This drawing shows the movement of the chromosomes toward the opposite poles of the cell during the _anaphase_ .
anaphase	2-81. After the anaphase is completed, the fourth step of mitosis, the <u>TELOPHASE</u> (TEL-o-phase) begins. The prefix TELO means end, so this phase is the _last_ step in the mitosis or reproduction of the cell.
last	2-82. TELOPHASE - I The fourth and final stage of mitosis, called the _telophase_, is the two-cell stage.
telophase	2-83. As the telophase begins, the chromosomes return to their normal size and are surrounded again by the nuclear _membrane_.

membranes	2-84. The process of mitosis is completed with the conclusion of its final phase, the telophase, with the formation of _two_ separate cells.

two

2-85.   Each of the two separate cells now contains _46_ (number) chromosomes.

TELOPHASE - II

46

2-86.   Let's review what you have learned about mitosis.

The objective of cell _~~division~~ mitosis_ is to produce new cells.

mitosis

2-87.

This drawing shows the cell while in the _resting_ phase, in other words, performing all its normal metabolic functions except reproduction.

resting

2-88.

A   B

The chromosomes thicken, the nuclear membrane disappears, and the chromosomes spread out around the nuclear area during the _prophase_.

prophase	2-89.

A   B

Drawing ___*A*___ shows the metaphase. Drawing B illustrates the ___*anaphase*___ .

A

anaphase

2-90.   The chromosomes become attached to fibre-like substances extending from the poles of the cell during the ___*metaphase*___ .

---

metaphase

2-91.   The anaphase is the stage during which the split chromosomes separate and move toward the ___*poles*___ of the cell.

---

poles

2-92   This is an illustration of the ___*telophase*___ or two-cell stage. Each new nucleus has ___*46*___ (number) chromosomes.

---

telophase

46

2-93.   The number of chromosomes is kept constant, 46 in each cell, because during the prophase each chromosome ___*splits*___ lengthwise.

---

splits

2-94.   Each body part has a different rate of speed at which its cells divide. Some body cells complete all five phases in 20 to 40 minutes, but this figure is not constant, since cells divide at ___*different*___ rates.

different	2-95	When you were first conceived, the total cell count of this "you" was <u>one</u>. It's amazing to consider that this one cell divided, and then the two new cells divided, and so on over and over through the process of _____ <u>mitosis</u> _____ until the final result was a human adult.
mitosis	2-96	If cells continued to divide without pausing to rest-- if each new cell proceeded to divide again as soon as it was "born"- - you can see that the new cells would become progressively smaller. It is the <u>resting</u> phase which allows the cells to _____ <u>grow</u> _____ to their normal size.
grow	2-97	The cells in the various body parts not only <u>divide</u> at different rates, but have different rates of _____ <u>growth</u> _____.
growth	2-98	Some interesting facts related to growth are: 1. Your size, <u>since conception,</u> has increased <u>26 billion times.</u> 2. If you had continued to grow at the same rate you did during the eighth month of pregnancy, your weight would be <u>two trillion times</u> the weight of the earth. (No response required - - Move to frame 99.)
	2-99	Obviously, you don't weigh more than the earth, and the reason you don't is that various factors have influenced the rate and nature of your cell _____ <u>growth</u> _____.
growth	2-100	Some growth factors are <u>inherited</u> in the _____ <u>chromosomes</u> _____ of the cell.

chromosomes	2-101	Nutritives - - foods - - influence growth, and so do certain <u>chemicals.</u> The vitamins and hormones and <u>DNA</u> are examples of growth - influencing chemicals.

DNA	2-102	You have developed a working understanding of the cell, its composition and how it reproduces and grows. Take a break before you continue. (No response required.)

	2-103	When a cell is not actively reproducing, it is said to be in the <u>resting</u> phase.

resting	2-104	The cell takes this opportunity to grow; it may also alter the <u>physical</u> and <u>chemical</u> properties of the cellular protoplasm during the <u>resting</u> phase.

resting	2-105	The altering of the chemical properties of protoplasm allows the cell to specialize. <u>SPECIALIZATION</u> of a cell's development takes place because the <u>chemical</u> and physical properties of its protoplasm are changed.

chemical	2-106	Some cells develop into skin, others into bone or other body parts. These are all examples of cell <u>specialization</u>

specialization	2-107	Once a cell begins to develop into a specific body part, it cannot change its course. When a cell's protoplasm is altered and the cell begins to develop into a specific stomach cell, it (will/will not) <u>will not</u> change and develop into a liver cell.

will not	2-108	If you recall, you developed from one cell that divided, and these new cells in turn divided, and so on until a many-celled being had developed. During the intervals between each division, certain cells became _specialized_ and developed into specific body parts.
specialized	2-109	A cell that has the potential of developing into a variety of body parts is termed MULTI-POTENTIAL (mul-ti-po-TEN-shal). Cells that can be either skin or bone, depending on how their protoplasm is altered, are examples of _multi-potential_ cells.
multi-potential	2-110	The body keeps some of these versatile _multi-potential_ cells in reserve at all times, in case of emergency.
multi-potential	2-111	All body cells go through the phases of reproduction, and then most go through a period of _specialization_ during which they become specific body parts.
specialization	2-112	As a cell becomes more specialized, it has less and less use for reproduction. The closer a group of cells comes to developing into, for instance, a completed nose, the less likely the individual cells are to _reproduce_
reproduce	2-113	Human cells rest first, then divide by the process of _mitosis_ , and then rest again.
mitosis	2-114	After the cell divides, the chemical and physical properties of its _protoplasm_ may alter, and the cell enters a period of _specialization._

protoplasm specialization	2-115	All body cells go through a cycle of (1) _mitosis_ and rest. Most cells then proceed to (2) _specialize_ until a particular body part is developed.
(1) mitosis (2) specialize	2-116	These specialized cells must be cemented together to give form and _structure_ to a body part.
structure	2-117	Between the cells is a cement-like substance called INTERCELLULAR (in-ter-CEL-u-ler) CEMENT. Arrow B points to the cytoplasm of the cell. Arrow A points to the intercellular _cement_.
cement	2-118	A brick wall when being erected has mortar placed between the bricks to make the bricks adhere to each other. Cells also have an adhesive substance, the _intercellular_ cement, that causes the cells to retain their relationship to each other.
intercellular	2-119	The inter-brick cement is supplied by the bricklayer, not made by the bricks. Cells, however, make their own adhesive, the _intercellular cement_.
intercellular cement	2-120	The intercellular cement is called a cell product. This indicates that intercellular cement is not itself a cell, but rather a cell _product_.

product	**2-121** The intercellular cement is only one of the many _products_ made by the cells.
products	**2-122** The space between the cells occupied by the intercellular cement is referred to as <u>INTERSTITIAL</u> (in-ter-STISH-al) <u>SPACE</u>, "interstitial" meaning "between one thing and another." The fluid which is also found in this space would be called _interstitial_ fluid.
interstitial	**2-123** Each cell is bathed on at least one side by _interstitial_ fluid.
interstitial	**2-124** You should not confuse the interstitial fluid with the fluid <u>inside</u> the cell, the <u>INTRA</u>cellular fluid. <u>Inter</u> means <u>between</u>, as in "intercellular" and "interstitial," and <u>intra</u> means <u>inside.</u>

INTERCELLULAR FLUID

Cytoplasm contains:

    a. intracellular fluid
       (Move to frame 126.)

    b. interstitial fluid
       (Move to frame 125.)

(b)	**2-125** You said that cytoplasm contains interstitial fluid. Not so! Cytoplasm is found inside the cell, not in the space outside the cell. (Move to frame 126.)

(a)	2-126	You said that cytoplasm contains intracellular fluid. Right! The prefix "intra" means inside; so the fluid within the cytoplasm, which is located inside the cell, is <u>intracellular</u> fluid. (Move to frame 127.)
	2-127	Body fluids may also be located in body tubing called <u>VESSELS</u> (VES-ls.) The adjective made from the noun <u>vessel</u> is <u>VASCULAR</u> (VAS-cu-lar.) Because these particular body fluids are transported <u>inside</u> the vessels, they are called _____*intra*_____ -vascular fluids.
intra	2-128	Blood <u>PLASMA</u> (PLAZ-ma) would be an example of _____*intra-vascular*_____ fluid.
intravascular	2-129	About 70% of the body weight of a human is accounted for by water. This 70% can be broken down into: 50% inside the cell, or (1) _____*intracellular*_____ fluid; 15% between the cells, or (2) _____*interstitial*_____ fluid; 5% in vessels, or (3) _____*intravascular*_____ fluid.
(1) intracellular (2) interstitial (3) intravascular	2-130	A body fluid that is <u>not</u> <u>intravascular</u> can be either _____*intracellular*_____ or interstitial.
intracellular	2-131	If a body fluid is not intracellular, it is either *interstitial* or intravascular.
interstitial	2-132	Intercellular cement is an example of: a. interstitial substance (Move to frame 134) b. intracellular substance (Move to frame 133.)

b.	2-133 You said that the intercellular cement was an example of <u>intracellular</u> substance.

The prefix <u>intra</u> means "inside." If the cement were intra-cellular, this would indicate that it is inside the cell, which is not the case.

(Move to frame 134) |
| a. | 2-134 Intercellular cement <u>is</u> an example of interstitial sub-stance.

Interstitial refers to any substance between the cells.

(No answer required. Move to frame 135) |

TISSUES

2-135 When the wall of a building is being erected, the arrangement of bricks and cement is not made simply at random. Groups of similar bricks are arranged in such a way that the wall will not fall down -- in other words, so that the wall will be able to per-form its proper <u>function</u> of support.

function	2-136 In order that a body may properly perform its function, <u>groups</u> of similar cells are held in place by the intercellular cement.
groups	2-137 Think of the confusion if eye cells were spread throughout the body. These eye cells would not be connected, and the func-tion of the eye could not be performed: in other words, you would be unable to <u>see</u>.
see	2-138 A group of cells that are similar have the same <u>form</u> and structure and also <u>function</u> in the same way.
function	2-139 TISSUES (TISH-uze) are the groups of cells with similar anatomy (form and structure) and <u>physiology</u> (function).

physiology	2-140 Muscle, nerve, bone, blood, and skin _tissues_ are composed of groups of similar cells.
tissues	2-141 Your finger is made of bone _tissue_ , muscle tissue, nerve _tissue_ and fat and skin tissue.
tissue tissue	2-142 The Rockettes that dance on the RCA stage in New York are like a tissue because they are somewhat similar structurally, and they dance or _function_ in unison.
function	2-143 If you were to study tissues, you would need to use a microscope. Such a study would be an examination of the _histology_ of the tissue.
histology	2-144 Histologically there are five basic tissues: 1. EPITHELIUM 2. SUPPORTING 3. MUSCLE 4. NERVE 5. CARDIAC (No response required. Move to frame 145.)

2-145   The <u>EPITHELIUM</u> (ep-i-THE-li-um) is a sheet-like tissue with one surface attached to your body and the other surface exposed. Skin is an example of epithelial tissue because one surface is attached to your body and the other surface is _exposed_ to the air.

UPPER DETACHED BORDER

EPITHELIUM TISSUE

LOWER ATTACHED BORDER

exposed	2-146

EPITHELIAL TISSUE

─UPPER SURFACE EXPOSED

─LOWER SURFACE *attached*.

(Fill in the blank)

attached	2-147 There are many hollow body spaces which are covered by this sheet-like *epithelial* tissue.

epithelial	2-148 Your coat can be compared to epithelial tissue. The inner coat lining is close to your body, and the outer surface is *exposed* to the weather.

exposed	2-149 Epithelial tissue <u>protects</u> the body and its parts in many ways. Skin, for example, prevents foreign substances from *entering* the body and causing infection.

entering	2-150 The body parts such as the stomach, kidneys and intestines are lined with epithelial tissue and are therefore *protected* by this tissue.

protected	2-151 Epithelial tissue forms <u>GLANDS,</u> groups of specialized cells used for the <u>elimination of waste</u>, the secretion of important fluids, and other jobs. The life function of *glands* *(excretion)* is carried out by the epithelial tissue.

excretion	2-152 Examples of specialized epithelial tissue cells which perform the function of excretion are the sweat ___glands___.

glands	2-153 Epithelial tissue also serves to <u>absorb</u> substances needed by the body. Food, for instance, is first broken down in the stomach, then ___absorbed___ by the lining of the intestine.
absorbed	2-154 Epithelial tissue also assists in performing the life function of <u>irritability</u> or the ___response___ of an organism to a stim-ulus.
response	2-155 If you have ever burned your tongue on food that was too hot, you have experienced an example of your tongue's epithelium performing the function of ___irritability___
irritability	2-156 Thirsting, smelling, seeing, hearing, and many other sensations are all irritability functions of the ___epithelium___.
epithelium	2-157 Finally, the epithelial tissues exhibit the life function of <u>reproduction</u>. If your skin is damaged, for instance, it is the epithelium which ___repairs___ it.

repairs	2-158	To sum up: the epithelial tissues can absorb and secrete important substances, exhibit the functions of irritability and excretion, protect and _repair_ parts of your body.

repair	2-159	You can always identify epithelial tissue because: 1. It is a sheet-like cover or (1) _lining_ for a body part; 2. One of the tissue's surfaces is (2) _attached_ to the body part, and the other surface is (3) _exposed_.

(1) lining (2) attached (3) exposed	2-160	Talented as it is, epithelial tissue is only one of the five basic tissues of the body. The others are: 2. supporting 3. muscle 4. nerve 5. cardiac The tissues in the bones which support your body are _supporting_ tissue.

supporting	2-161	Tissues that help you move and lift objects are _muscle_ tissues.

muscle	2-162	Nerve tissues share with epithelial tissues the job of helping you respond to sensations, in other words, they help in the life function of _irritability_. Cardiac tissues are found in the muscles of the heart.

irritability	2-163	The five body tissues are: 1. the tissues that cover or line body parts, (1) _epithelial_ tissue; 2. tissue such as bone tissue, (2) _supporting_ tissue. 3. (3) _muscle_ tissues that help you work and move; 4. (4) _nerve_ tissues that help you respond. 5. (5) _cardiac_ tissues in the heart.

(1) epithelial (2) supporting (3) muscle (4) nerve (5) cardiac	2-164	Take a break. (No response required.)

	2-165	Let's review what you've learned in this unit. The small "building blocks" of the human body are specialized forms of protoplasm called ___cells___.

---

cells	2-166	Fill in the blanks: 1. _cell membrane_ 2. _cytoplasm_ 3. _nuclear membrane_ (nucleus) 4. _chromosomes_ 5. _nucleoplasm_

---

(1) cell membrane (2) cytoplasm (3) nucleus (4) chromosomes (5) nucleoplasm	2-167	The cell nucleus influences cellular _reproduction_.

---

reproduction	2-168	Cellular metabolism is carried out in the cell "workhouse", the _cytoplasm_.

---

cytoplasm	2-169	Another name for the normal process of cell reproduction in the human body is _mitosis_.

mitosis	2-170 Cell division or mitosis is divided into four stages or phases. The nuclear membrane disappears and the chromosomes split during the _prophase_.
prophase	2-171 During the metaphase the chromosomes become attached to delicate fibre-like substances which extend to the opposite cell _poles_.
poles	2-172 During the anaphase the split chromosomes move toward the poles of the cell. There are now _46_ chromosomes on each side of the cell.
46	2-173 The last phase of division, when two distinct and separate cells are formed, is the _telophase_.
telophase	2-174 Cells divide, then pause to grow and _specialize_ into different forms before reproducing again.
specialize	2-175 A multipotential cell is one that can grow into one of _many_ body parts.
many	2-176 Cells are held together by an intersititial substance called _intercellular_ cement.
intercellular	

CHAPTER 3

Human Tissues

3-1.   Cells that have a similar anatomy and physiology are grouped together and classified as body _____*tissues*_____ .

---

tissues

3-2.   An example of a tissue would be the _____*epithelium*_____ , which lines and protects body parts.

---

epithelium

3-3.   The epithelium is one of the five basic body tissues. The other four basic tissues are _____*nerve*_____ , _____*muscle*_____ , supporting, and cardiac tissues.

---

muscle, nerve
(either order)

3-4.   Supporting tissues are grouped together because of <u>functional similarities</u>, how they act, rather than structural similarities. Although they may not look alike, _____*supporting*_____ tissues do <u>function</u> alike.

---

supporting

3-5.   You know that the epithelial tissue protects and repairs damaged body areas. Supporting tissue also functions as a _____*protective*_____ and repair agency.

protective	3-6.	When you think of supporting tissue you should remember:

Supporting tissues are more similar
in ___*function*___ than they are in structure.

function	3-7.	Supporting tissues support, ___*protect*___ and ___*repair*___ the body.

repair, protect (either order)	3-8.	Epithelial tissue has <u>small amounts</u> of intercellular cement between the cells. Supporting tissue, however, contains much more intercellular substance than does epithelial tissue. In some types of supporting tissue--bone, for example -- the ___*intercellular*___ substances are more abundant than the cells.

intercellular	3-9.	All cells are bathed on at least one side by interstitial fluid. The epithelial tissue cells receive support from the ___*intercellular*___ cement which occupies the interstitial space.

intercellular	3-10.	The interstitial substance of supporting tissue is called the tissue <u>MATRIX</u> (MA-triks). For example, the calcium which surrounds bone cells is part of the supporting tissue ___*matrix*___ .

matrix	3-11.	Supporting Tissue <u>MATRIX</u> (interstitial substance)

<u>FIBERS</u>   <u>GROUND SUBSTANCE</u>

The supporting tissue matrix is composed of <u>FIBERS</u> and ___*ground substance*___ .

ground substance	3-12. Epithelial tissue cells are surrounded by (1) _interstitial_ fluid. Supporting tissue cells are surrounded by a tissue matrix which is composed of (2) _fibers_ and (3) _ground_ substance.
(1) interstitial (2) fibers (3) ground	3-13. The ground substance is semifluid or solid, depending upon the type of supporting tissue. Bone would be an example of a structure whose ground substance is _solid_.
solid	3-14. Ground substance is often amorphous, like gelatin; in other words, unlike the body, it has no definite _form_ or structure.
form	3-15. If you mixed some liquid gelatin and added to this a cup of fruit and a cup of short pieces of string, you would have, when the mixture set, a good imitation of a _supporting_ tissue.
supporting	3-16. Let's examine the gelatin analogy. The pieces of fruit embedded in the gelatine would be the: a. cells of the tissue (move to frame 18) b. ground substance of the tissue (move to frame 17)
(b)	3-17. You said that the pieces of fruit would be the ground substance. Do you recall in frames 9-12 (review if necessary) that the ground substance <u>surrounds</u> the cells? Your choice was wrong. Move to frame 18 for more information.
(a)	3-18. You said that the pieces of fruit would be the cells. Right! The ground substance, or gelatine in this example, surrounds the cells (pieces of fruit) in a semiliquid to solid state. Move to frame 19.

3-19.   You have accounted for the gelatine -- the ground substance -- and the fruit -- the cells. The pieces of string would be the second component of a tissue matrix, the _fibers_.

---

fibers

3-20.   The ground substance of the tissue matrix can be in either a semifluid or a _solid_ state.

---

solid

3-21.   Some books indicate that supporting and CONNECTIVE tissues are one and the same thing. You will recall, however, that tissues are classified as either supporting, muscle, nerve, epithelial or cardiac tissues.

(No response required)

3-22.   The confusion between supporting and connective tissue arises from the fact that _connective_ tissue is only one type of supporting tissue.

---

connective

3-23.   The cells of connective tissue are called FIBROBLASTS (FI-bro-blasts). The suffix,-blast, means to sprout or germinate. The word fibroblast indicates that the connective tissue cells sprout or form _fibers_.

---

fibers

3-24.   It is interesting to note that the fibers formed by the connective tissue cells or _fibroblast_ are not attached to the cells.

---

fibroblasts

3-25.   Fibers formed by the fibroblasts are cell _products_ similar to intercellular cement.

59

products	3-26. Metabolism and other cell activities are carried on by the _fibroblast_ of the connective tissue.
fibroblasts	3-27. Let's review what you have learned. Supporting tissues are groups of cells alike in (1) _function_ but not necessarily in (2) _structure_.

(1) function
(2) structure
	3-28. Supporting tissue is
	
	  a. cells matrix (move to frame 29.)
	
	  b. cells ground substance (Move to frame 30.)

(a)	3-29. You said that supporting tissue is cells and matrix. Right. (Move to frame 31.)
(b)	3-30. You said that supporting tissue is cells and ground substance. Not so. Review frames 11 – 20 and then return to frame 28.
	3-31. Ground substance is produced by the _fibroblast_ of connective tissue.
fibroblasts	3-32. The fibers that are germinated or made to grow by the fibroblasts determine what type of tissue is present. For example, COLLAGENOUS (kol-AG-in-us) connective tissue is so called because the protein collagen is the chief ingredient of its _fiber_.

60

fibers	3-33. Collagenous connective tissue is made of <u>wavy</u> fibers.

ONE FIBER

~~~~~~~~~~~~~~~~~~~~~

</div>

This is a fiber of _____collagenous_____ connective tissue.

---

collagenous	3-34. These wavy fibers are composed, in turn, of wavy layers of <u>FIBRILS</u> (FI-brils). A <u>fibril</u> is simply a very small fiber.

Look back at the drawing in frame 33. There are three fibrils in this fiber of _____collagenous_____ connective tissue.

---

collagenous	3-35. The size of a collagenous fiber is determined by the number of _____fibrils_____ that it contains.

---

fibrils	3-36. Fibers have a cement that surrounds the fibrils somewhat like the insulation on electrical cables that holds strands of wire together.

<div align="center">

ONE FIBER   CEMENT

fibril
(FILL IN)

CEMENT

</div>

---

fibril	3-37. What might be the advantage of wavy collagenous fibers?

a. to provide the tissues with great strength and elasticity. (Move to frame 38.)

b. to prevent fibroblasts from leaving the tissue. (Move to frame 39.)

(a)	3-38. You said the wavy collagenous fibers provide great strength and elasticity. Right.
	
	    The fibrils can be stretched until straightened and the fiber then becomes very resistant to any further change. If the fiber has not been stretched so much that it is damaged, it will return to its original position when released.
	
	    (Move to frame 40.)

(b)	3-39. You said that the wavy collagenous fibers prevent the fibroblasts from leaving the tissue. Not so. Go back and review frame 24, and then reread frame 37.
	3-40. Collagenous fibers are extremely tough. This quality of toughness of collagenous fibers makes this connective tissue resistant to _damage_ (your own words).
damage	3-41. When fat and other impurities have been removed from certain collagenous tissues, the fibers can be used as surgical gut to repair injured tissues. (No response required. Move to frame 42.)
	3-42. There are other connective tissue fibers whose most important component is the protein elastin, and these fibers are called ELASTIC fibers. The name elastic tells you that these _elastic fibers_ are more flexible than collagenous fibers.
fibers	3-43. The most resilient connective tissue fibers are the _elastic_ fibers.
elastic	3-44. Both elastic and collagenous fibers are embedded in the amorphous _ground_ _substance_ of the tissue matrix.

ground substance	**3-45.** The consistency of the tissue matrix varies with <u>ACTIVITY</u>, <u>AGE</u>, and the general <u>HEALTH</u> of the body. The tissue matrix of a healthy <u>child</u> may vary considerably from that of a healthy ___*adult*___.
adult	**3-46.** Connective tissue has <u>three</u> types of fibers, elastic, <u>RETICULAR</u> (re-TIK-u-ler), and ___*collagenous*___.
collagenous	**3-47.** In some texts you will find collagenous and reticular fibers classed as the same fiber type, because reticular fibers also contain the protein collagen. This course, however, classes ___*reticular*___ and collagenous fibers separately.
reticular	**3-48.** Reticular connective <u>tissue</u> is composed of a mesh of fine ___*reticular*___ fibers.
reticular	**3-49.** In some locations, reticular connective tissue is called the <u>BASEMENT MEMBRANE</u> because it acts as a base or foundation for the epithelium. In other locations it provides the framework for various organs, the liver, for instance, is constructed around a meshwork of ___*reticular*___ tissue.
reticular	**3-50.** Don't become confused. Reticular <u>tissue</u> is so-called because it consists of reticular fibers, which may be present in many tissues other than reticular connective tissue. (No response required. Move to frame 51.)
	3-51. Another type of connective tissue containing reticular fibers is <u>ADIPOSE</u> (AD-i-pose) or simply <u>fat</u> ___*tissue*___.

tissue	3-52. Fat storage and the production of fat cells are the main function of _____*fat*_____ tissue.
fat	3-53. The state of nutrition determines how much fat or _____*adipose*_____ tissue is present in the body.
adipose	3-54. No one is literally a "fat-head," since the brain, like the eyes and the lungs, is unable to store _____*fat*_____ .
fat	3-55. Adipose cells are supported by a mesh of _____*reticular fibers*_____ .
reticular fibers	3-56 Concrete is often reinforced by the addition of a wire mesh. Adipose tissue is _____*reinforced*_____ by the network of reticular fibers. RETICULAR FIBERS FAT CELLS
reinforced	3-57. Adipose tissue is a good body packing agent and shock absorber and acts also as an _____*insulator*_____ against cold.

insulator	3-58. Pause for a moment and review what you've learned. Connective tissue has three kinds of <u>fibers</u>: the tough, resilient (1) _collagenous_ fibers; the even more resilient (2) _elastic_ fibers; and the mesh-forming (3) _reticular_ fibers.
(1) collagenous (2) elastic (3) reticular	3-59. A collagenous fiber is composed of wavy _fibrils_.
fibrils	3-60. The type of connective <u>tissue</u> present is determined in part by the type of fiber in the tissue. Collagenous tissue, for example, contains the protein collagen, which is the chief ingredient of its tough _fiber_.
fibers	3-61. Collagen is also an element of the somewhat finer fibers which form the mesh-like _reticular_ tissue, which is sometimes called the basement _membrane_.
reticular membrane	3-62. So far you have read about connective tissues that have been loosely arranged. Connective tissue can also be dense, so dense that the matrix contains more _fibers_ than ground substance.
fibers	3-63. Flex your foot upward, and then reach down and feel the large cordlike structure that connects your calf muscle to your heel. You are feeling the <u>ACHILLES</u> (a-KIL-ez) <u>TENDON</u> (TEN-dun), a good example of _____ connective tissue. ACHILLES TENDON

dense	3-64. The Achilles tendon is an example of connective tissue in which the _____ are especially numerous and closely packed.
fibers	3-65. Tendons are composed predominantly of wavy collagenous fibers. If you recall, these fibers could be stretched to a point beyond which the tissue would be _____.
damaged (or injured)	3-66. Tendons are stronger, in some respects, than a bone of equal size. For example, the _____ fibers of a 1/4 " tendon can sustain a 1,000 pound load!
collagenous	3-67. Watch a human or a dog jump, and you will see that tendons allow muscles to contract and then to recoil to their original position. You should remember that tendons attach _____ to _____.
muscles bones	3-68. Another dense connective tissue that may be cord-like is the LIGAMENT (LIG-a-ment). Ligaments use both elastic and collagenous fibers to connect bones together. Tendons attach bones to muscles. _____, on the other hand, help connect bones to each other.
ligaments	3-69. Feel the lower inch of your nose. Wiggle it about. Is it connected to the upper bony structure? This material under the skin is (a) connective tissue (move to frame 71) (b) soft bone (move to frame 70)
(b)	3-70. You said that the material was soft bone. Not bad! The material is called CARTILAGE (KAR-ti-lij). Cartilage is a dense connective tissue which is firm, like plastic, but not hard like a bone. In infancy, cartilage occupies some areas which later become bone. (move to frame 72)

(a)	3-71. You said that the material was connective tissue. Very good! This connective tissue is called CARTILAGE (KAR-ti-lij). Cartilage is a dense connective tissue which is <u>firm</u>, like plastic. In infancy, cartilage occupies some areas which later become bone.
	3-72. The dense connective tissue in your ear is another example of _cartilage_ .
cartilage	3-73. As you walk, jump, or run, your bones do not rattle against one another because of the pads of _cartilage_ between them.

cartilage	3-74. It is important for you to note that dense connective tissue does not permit blood vessels or nerves to pass through it easily. If tissues such as cartilage, tendons, and ligaments are damaged, they will
	
	    a. repair slowly
	    b. repair fast
	
	Select either <u>a</u> or <u>b</u> and move to frame 75.

a.	3-75. Tissues with a lower blood supply than is average for the rest of the body will heal more _____ than the other parts of the body.
slowly	3-76. As you become older, the cartilage between the bones becomes, for reasons unknown, increasingly inflexible, and the joints become less and less mobile. Cartilage can be converted to _bone_ _____ early or late in your life.

bone	3-77.	Supporting tissue, you recall, is designed to _protect_ and support the body.

protect	3-78.	The final supporting tissue you will study supports your body and protects such vital organs as your brain and your heart. This densest of supporting tissue is _bone_.

bone	3-79.	Bone, like all other supporting tissue, contains cells and a matrix of _ground_ _substance_ and _fibers_.

ground substance fibers	3-80.	There is one distinguishing characteristic of bone that makes it different from other supporting tissue. The ground substance of bone contains a <u>high content</u> of inorganic calcium compounds which make the bone tissue _rigid_.

rigid or hard	3-81.	Bone tissue <u>is</u> <u>not</u> a solid structure but has many internal <u>canals</u>. The canals allow _fluids_ to pass through the bone and supply nutrition to the bone cells.

fluids	3-82.	The other supporting tissue cells you learned were the fiber-producing _fibroblasts_. Bone cells are called <u>OSTEOCYTES</u> (OS-te-o-sites).

fibroblasts	3-83. This cross section of part of a bone shows three bone cells or *osteocytes* .
osteocytes	3-84. The bone cells or *osteocytes* are situated in small <u>cavities</u> in the bone.

osteocytes	3-85. Look at the illustration in frame 83.
	     The cell and its parts are drawn in red; the portion drawn in black is the bone matrix. Note that:
	     1. The osteocyte membranes spread out like roots.
	     2. The bony rings are similar to the growth rings of a tree.
	
	     (No response required. Move to frame 86.)

	3-86. Bone tissue is not solid. In addition to cavities for the osteocytes, it contains canals which allow the osteocytes to receive *nutrition* .
nutrition	3-87. Nutrients pass through the canals themselves or are conducted by <u>blood</u> <u>vessels</u> within the *canals* .
canals	3-88. The canals pass <u>up</u> and <u>down</u> through the bone and <u>cross</u> the rings of bone that contain the *osteocyte* cavities.

osteocyte	3-89. In this illustration the matrix is drawn in red, and the _____ in black.

canals	3-90. But why is such an elaborate system of canals needed in bone tissue?
	    (a) bone is so dense that the nutrients could not pass from cell to cell easily.
	        (move to frame 92)
	    (b) the tissue length requires a canal system.
	        (move to frame 91)

(b)	3-91. You said that the tissue length requires an elaborate canal system. Not so.
	
	    Collagenous and adipose tissue may be equally long; yet they need no such system.
	
	    (move to frame 89)

(a)	3-92. You said that the bone tissue was so dense that a canal system was required. Right!
	
	    The nutrients from the blood vessels could not easily penetrate the inorganic _____ deposits; as a result, the osteocytes would die.
	
	    (move to frame 93)

calcium	3-93. If a bone is completely burned, the white calcium and other inorganic _____ _____ survive, but the fibers are destroyed.
ground substance	3-94. The fibers in the matrix of one growth layer of the bone interwine with the fibers of other layers to provide greater bone _____.

strength	**3-95.** The next time you have a chicken or turkey leg, place the bone in a jar and cover it with a vinegar solution. In a few days, you will be able to tie this customarily brittle bone into a knot. Why? a. The vinegar made the bone more elastic. (Move to frame 97.) b. The vinegar dissolved the calcium compounds in the bone. (Move to frame 96.)
(b)	**3-96.** You said that the vinegar dissolved the calcium in the bone. Right! When the calcium compounds were dissolved, all that was left were the _____ of the matrix. (Move to frame 98)
(a) fibers	**3-97.** You said that the vinegar made the bone more elastic. Well, yes and no. Yes, the bone was more elastic. But the vinegar <u>did something</u> to the bone to make it more elastic. (Move to frame 96)
	3-98. The calcium compounds that are stored in bone are in a constant state of flux. While some calcium is being deposited, other portions are being removed for use elsewhere in the body. The portions being transported <u>out</u>, like the nutrients which are transported <u>in</u>, are carried by the _____ _____.
blood vessels	**3-99.** Bones also manufacture and store <u>MARROW</u> (MAR-o). There are two types of marrow, red marrow and yellow. Your bones contain both red and yellow _____.
marrow	**3-100.** Marrow is responsible for the production of blood cells. If the marrow is diseased, a bone and/or _____ disease could develop.

blood	3-101. Bone and bone marrow are separate entities. Marrow is a cell _____ of bone tissue.

product	3-102. Bones have still another function in the body – they are good levers. Aside from housing marrow and providing a mooring place for muscles, the four functions of bones are that they:
	
	    1. protect the body.
	    2. (1) _____ the body.
	    3. act as (2) _____ allowing motion.
	    4. store and release (3) _____ .

1. support 2. levers 3. calcium	3-103. At the beginning of this unit you learned about <u>loosely-arranged</u> connective tissues – the collagenous, reticular and elastic tissues. Let's review what you have read about bone and cartilage or _____ connective tissue.
dense	3-104. What makes connective tissue either more or less dense is the proportion of (1) _____ to (2) _____ _____ in the matrix. The higher the proportion of (3) _____ , the denser the tissue.
1. fibers 2. ground substance 3. fibers	3-105. A cord-like connective tissue which attaches bones to muscles is a _____ .
tendon	3-106. A connective tissue – also sometimes cord-like – which attaches bones to <u>each other</u> is a _____ . A tissue which is firm and flexible, like plastic, is _____ .

ligament cartilage	3-107. Because of their density, these tissues allow proportionately less blood to pass through them than the rest of the body and therefore heal more _____ .
slowly	3-108. The densest supporting tissue in the body is _____ . It derives its hardness or rigidity from its high _____ content.
bone calcium	3-109. Bones are not solid but contain cavities for both the _____ and _____ which allow nutrients to enter.
osteocytes canals	3-110. Bones also manufacture and store _____ , which is in turn responsible for the production of certain _____ cells.
marrow blood	3-111. A break is in order. Rest a bit before you continue.
	3-112. The next type of tissue to be studied after epithelial and supporting is the "meat" or "flesh" of the body, the _____ tissue.
muscle	3-113. Muscle tissue differs from other tissues because a single muscle fiber is a <u>muscle</u> <u>cell</u>. Epithelial tissue has no fibers; however, each muscle cell is a _____ _____ .

73

muscle fiber	3-114. The primary function of muscle tissue is to do <u>work</u> by first contracting and then _____ again.

---

expanding	3-115.

MUSCLE CELLS

NUCLEI

1. STRIATED (STRI-a-tid)   2. SMOOTH

Muscle cells are classified as being either striated, meaning striped, or _____.

---

smooth	3-116. Look at the illustration in frame 115.

The cross-striped cell is a _____ muscle cell. The unstriped cell is a _____ muscle cell. |

---

striated smooth	3-117 "Striated" and "smooth" refer not to the <u>function</u> of a muscle cell, but to the muscle's _____.

---

structure	3-118. Muscles can also be further classified as <u>voluntary</u> or <u>involuntary</u>. This classification refers to how the muscle cells _____.

function	3-119. A muscle such as the large muscle (biceps) in your arm is <u>striated</u>, and because you have control over the actions of this muscle it is referred to as a <u>striated</u> (voluntary / involuntary) _____ muscle.
voluntary	3-120. You do not have any control over the smooth muscles of your body. Therefore, there <u>are</u> <u>no</u> smooth _____ muscles.
voluntary	3-121. All smooth muscles of the body are _____.
involuntary	3-122. The striated muscles are all voluntary with one exception, the heart, which is cardiac muscle. Because you have no control over your heart, it is an _____ muscle.
involuntary	3-123. Skeletal muscles, like those in your arms and legs, are striated. Your leg muscles are therefore striated _____ muscles.
voluntary	3-124. Remember, there are _____ <u>smooth</u> voluntary muscles and but one involuntary <u>striated</u> muscle, the _____ muscle.
no heart	3-125. Smooth muscle cells have simple cellular structures with tapering ends, a central _____ and many small organoids.

nucleus	3-126. The organoids of smooth muscles are embedded in the cell's cytoplasm, which in muscle cells is called the SARCOPLASM (SAR-ko-plazm). The organoids called MYOFIBRILS (MĪ-o-fi-brilz) are embedded in the _____ of the muscle cell.

---

sarcoplasm

3-127. The <u>contracting</u> element of the muscle are the small organoids called _____.

---

myofibrils

3-128. <u>SMOOTH MUSCLE CELL</u>

Fill in the blanks.

(1.) _____
(2.) _____
(3.) _____

---

1. nucleus
2. myofibril
3. sarcoplasm

3-129. You recall that muscle tissue does not have separate cells and fibers. The muscle fiber <u>is</u> the cell and includes the nucleus, the _____, and the myofibrils which are the _____ element of the muscle.

---

sarcoplasm
contracting

3-130.

SMOOTH MUSCLE TISSUE

Each muscle fiber is separate and is contained within a delicate sheet of <u>connective</u> _____.

tissue	3-131. The bundles of separate _____ are then bound together by more connective tissue.
fibers	3-132. If an impulse is transmitted to one fiber in the <u>bundle,</u> it in turn transmits the stimulus to every fiber in the <u>bundle,</u> and the myofibrils within each cell react by _____.
contracting	3-133. The skeletal or striated muscle fibers also contain cytoplasm called _____ and the contracting elements, the _____.
sarcoplasm myofibrils	3-134.

STRIATED MUSCLE FIBER
(LONGITUDINAL SECTION)

SARCOPLASM   MYOFIBRILS

STRIATIONS   NUCLEI

The dark and light bands that give this muscle its striped or striated appearance are due to the structure and arrangement of the _____.

myofibrils	3-135. The cells of striated muscles contain not just one nucleus but several, making them _____ - nucleated cells.
multi	3-136. Like smooth muscle fibers, the individual striated fibers are surrounded by _____ tissue. Groups of fibers are bound together to make _____ of fibers.

connective bundles	3-137. Striated muscles respond in much the same way that smooth muscles do. The impulse is transmitted from fiber to fiber so that the entire _____ contracts at once.
muscle	3-138. If sarcoplasm is present in proportionately large quantities, the muscle is less susceptible to fatigue. The muscle of your heart -- which is constantly at work -- contains (more/less) _____ sarcoplasm than do your cheek muscles.
more	3-139. Your arm and hand are easily <u>fatigued</u>, but they can <u>move quickly</u>. Quick movement in muscles indicates that a muscle has (more/less) _____ sarcoplasm in its fibers.
less	3-140. A low level of sarcoplasm within the muscle cells makes the muscle _____ susceptible to fatigue. A high level of muscle cell sarcoplasm _____ the speed of muscular movement.
more reduces	3-141. When a muscle contracts, work is accomplished. The <u>energy</u> required to do this _____ is stored within the muscle.
work	3-142. Muscle cells store phosphate compounds, referred to as <u>ADP</u> or <u>ADENOSINE</u> (a-DEN-o-sin) diphosphate and <u>ATP</u> or <u>ADENOSINE</u> triphosphate. ATP breaks down to ADP, and energy is released. Contraction, or _____, is accomplished.
work	3-143. The energy supply for muscular contraction is produced when the _____ molecules breaks down to ADP.

ATP	3-144	Let's review some of what you've been learning. 1. Connective tissue has distinct cells and fibers. 2. The epithelium has no (1) _____ . 3. In muscle tissue, the individual muscle (2)_____ <u>are</u> the (3) _____ .
(1) fibers (2) cells (3) fibers (2 and 3-- either order)	3-145	All muscles can be classified either as voluntary or involuntary. How many <u>smooth</u> voluntary muscles are there in your body? (none, few, many) _____ .
none	3-146	Heart muscle is striated; you would classify this as _____ striated muscle.
involuntary	3-147	All striated muscles except _____ muscle can also be referred to as skeletal muscles.
heart	3-148	Fill in the blanks on this illustration of a (1) _____ muscle cell. NUCLEUS 2. _____ 3. _____
(1) smooth (2) sarcoplasm (3) myofibril	3-149	Muscle cells are enclosed in separate sheets of tissue and combined into bundles which are also enclosed in _____ _____ .
connective tissue	3-150	Striated muscles contain proportionately less muscle cytoplasm than smooth muscles. Therefore striated muscles both contract and tire _____ . a. less quickly than smooth muscles. (Move to frame 152) b. more quickly than smooth muscles. (Move to frame 151)

(b)	3-151	You're right! Because striated muscles have less sarcoplasm, they both contract and tire more quickly than smooth muscles. (Move to frame 153.)
(a)	3-152	You said that striated muscles contract and tire less quickly than smooth muscles. Not so! Striated muscles have less sarcoplasm than smooth muscles, and the less sarcoplasm present in a muscle, the more quickly it can contract and the more quickly it tires. (Move to frame 153).
	3-153	NERVE TISSUE The ability of a group of cells to respond to a stimulus, the life function _____, is one of the most fascinating of life functions.
irritability	3-154	When the body or one of its parts responds to a stimulus, it is the _____ cells that are responsible for transporting the impulse from the cells to sections of the body.
nerve	3-155	Remember that nervous tissue is primarily a receiver and conductor. A muscle responds when the nervous tissue receives a stimulus and _____ the impulse to that muscle.
conducts	3-156	All protoplasm exhibits the qualities of irritability and conductivity to some degree, but these qualities are most highly developed in _____ _____.
nerve cells or nervous tissue	3-157	A nerve cell is called a NEURON (NU-ron). The special functions of neurons are to receive _____ and to conduct impulses.

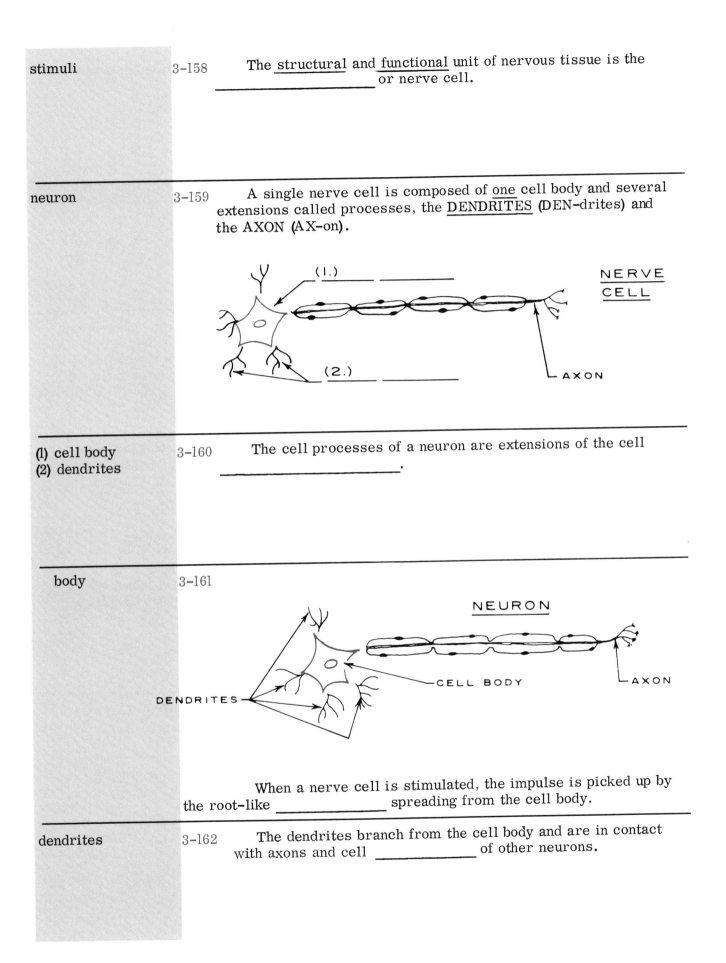

stimuli

3-158 The <u>structural</u> and <u>functional</u> unit of nervous tissue is the _____ or nerve cell.

neuron

3-159 A single nerve cell is composed of <u>one</u> cell body and several extensions called processes, the <u>DENDRITES</u> (DEN-drites) and the AXON (AX-on).

NERVE CELL

(1.) _____

(2.) _____

AXON

(1) cell body
(2) dendrites

3-160 The cell processes of a neuron are extensions of the cell _____.

body

3-161

NEURON

DENDRITES

CELL BODY

AXON

When a nerve cell is stimulated, the impulse is picked up by the root-like _____ spreading from the cell body.

dendrites

3-162 The dendrites branch from the cell body and are in contact with axons and cell _____ of other neurons.

bodies	3-163 The process of the neuron that carries an impulse <u>away</u> from the cell body is the _____. The processes of the neuron that transmit the impulse toward the cell body are the _____.
axon dendrites	3-164 Neurons may have many dendrites; however, they have but one cell body and one _____.
axon	3-165 The impulses that travel through a neuron are classed as either <u>AFFERENT</u> (AF-er-ent) or <u>EFFERENT</u> (EF-er-ent). When you come <u>into</u> a theater, you are <u>admitted</u> by the ticket-taker; when you go <u>out</u>, you leave by the <u>exit</u>. The prefix <u>ex</u> means out, and the prefix <u>ad</u> means _____.
in or into	3-166 An <u>AFFERENT</u> (AF-er-ent) impulse is one that travels <u>toward</u> the cell body of a neuron. Afferent impulses would travel via the _____ of a neuron.
dendrites	3-167 An <u>EFFERENT</u> (EF-er-ent) impulse is one that travels <u>away</u> from the cell body of a neuron. Efferent impulses would travel via the _____ of a neuron.
axon	3-168 Dendrites transmit _____ impulses. Axons transmit _____ impulses.
afferent efferent	3-169 The axon, drawn here in red, is enclosed in protective covers called _____.

sheaths	3-170 Depending upon the location and function of an axon there may be <u>more</u> than one _____ covering it.
sheath	3-171 Two types of sheaths are identifiable, the <u>MYELIN</u> (MY-e-lin) and the <u>NEUROLEMMA</u> (nu-ro-LEM-a). . An axon may have either one type of sheath or the other, or both. When <u>both</u> are present, the myelin is always the inner sheath, and the _____ the outer sheath.
neurolemma	3-172 If an axon is enclosed in <u>two</u> sheaths, the neurolemmal sheath is always _____ the myelin.
outside (of)	3-173 The neurolemmal sheath is cellular in nature. The _____ sheath is non-cellular.
myelin	3-174 The myelin sheath is a <u>non-living</u> fatty substance. When the neurolemma is present, it produces the myelin; but when it is not present, we do not yet know how the non-living_____ sheath is produced.
myelin	3-175 The <u>WHITE MATTER</u> of the brain is a non-cellular material composed mostly of (neurolemma/myelin) _____.
myelin	3-176 The <u>GRAY MATTER</u> of the brain is composed of nerve cells and <u>cellular</u> material which is _____.

neurolemma	**3-177** The neurolemma has another function besides that of protection: it plays an active role in <u>cell repair</u>. Fully developed nerve cells cannot reproduce, but the sheath-covered cell process, the _____, can be <u>repaired</u>.
axon	**3-178** For example, if an axon is cut, the part that is separated from the cell body degenerates, but the sheath survives. If the separated neurolemmal sheath and the sheath which is still attached heal together again, a new _____ grows out into the reconnected sheath.
axon	**3-179** You will discover later in this course that certain processes of the nervous <u>system</u> cannot be repaired. The action of the neurolemma referred to in frames 177-178 applies to many but <u>not all</u> processes.
	3-180 Let's review some of what you have learned about nervous tissue. A neuron is the _____ and _____ unit of nervous tissue.
structural functional	**3-181** Neurons have two main functions, to _____ stimuli and _____ impulses.
receive conduct	**3-182**

NEURON

Label the parts drawn in red.

(1) dendrite
(2) axon
	3-183	An afferent impulse is conducted by
		
		    a. an axon (Move to frame 184)
		
		    b. dendrites (Move to frame 185)

(a)	3-184	You said that an afferent impulse is conducted by an axon. Not so.
		
		    Remember, <u>afferent</u> impulses are those travelling <u>toward</u> the cell body, and axons carry impulses <u>away</u> from the cell body.
		
		    (Move to frame 186)

(b)	3-185	You said that an afferent impulse is conducted by dendrites. Right! Now move to frame 186.
	3-186	When cell axons are protected by two sheaths, the cellular _____ sheath is the outer insulation, and the non-cellular _____ sheath is the inner insulation.
neurolemma myelin	3-187	If you were to describe the myelin, you would say that it was a _____ fatty substance. An example of myelin is the _____ matter of the brain.
non-cellular white	3-188	Although some neurons can be repaired, mature neurons can <u>not</u> _____.
reproduce	3-189	The repairing of an axon can take place if the detached _____ sheath heals and a new protective sheath is formed.

neurolemmal	3-190 You have discovered that neurons transmit impulses. You are called, and you react; you are physically pushed, and again you react. How does a stimulus get from your ear or skin to the central computer, your brain? (No response required. Move to frame 191.)
	3-191 When individual neurons are placed end to end, they form extensive neural pathways. When you are called, the receptor neurons in your ears send an impulse to other _____ which finally transmit the message to the brain.
neurons	3-192 It is important to remember that the impulse travels from the <u>axon</u> of one neuron either to the _____ or directly to the cell body of the next neuron.
dendrites	3-193 Note that there is <u>never</u> a structural connection between the _____ of one neuron and the dendrites of another neuron.
axon	3-194 The association between the axon of one neuron and the dendrites or cell body of another is called a <u>SYNAPSE</u> (si-NAPS). The synapse is a <u>functional</u> connection rather than a _____ connection.

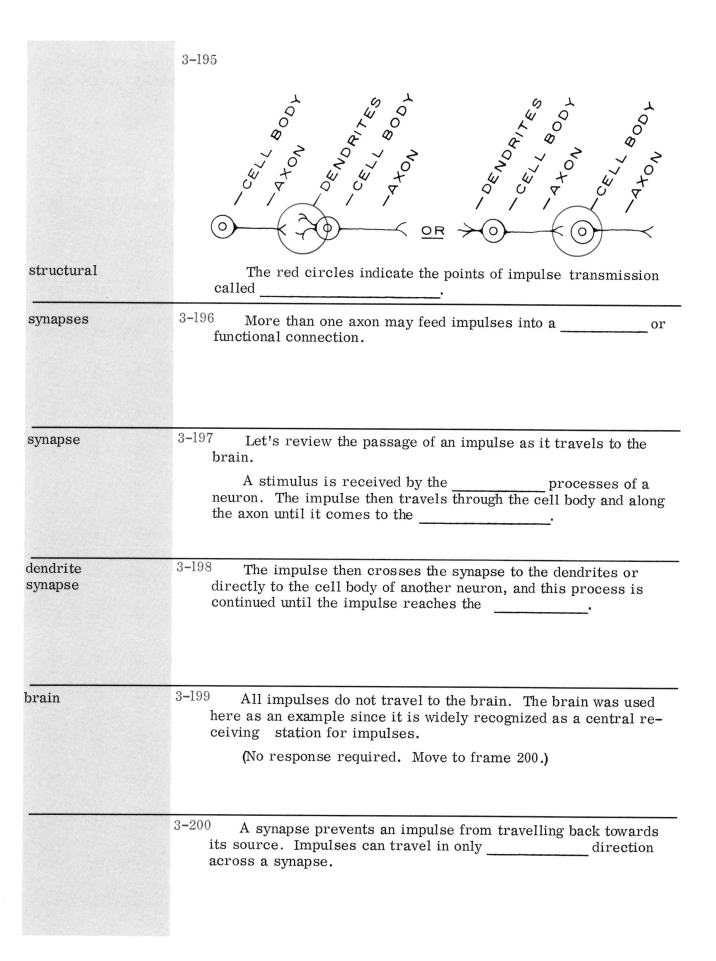

3-195

The red circles indicate the points of impulse transmission called _____.

structural

---

synapses

3-196   More than one axon may feed impulses into a _____ or functional connection.

---

synapse

3-197   Let's review the passage of an impulse as it travels to the brain.

A stimulus is received by the _____ processes of a neuron. The impulse then travels through the cell body and along the axon until it comes to the _____.

---

dendrite
synapse

3-198   The impulse then crosses the synapse to the dendrites or directly to the cell body of another neuron, and this process is continued until the impulse reaches the _____.

---

brain

3-199   All impulses do not travel to the brain. The brain was used here as an example since it is widely recognized as a central receiving station for impulses.

(No response required. Move to frame 200.)

3-200   A synapse prevents an impulse from travelling back towards its source. Impulses can travel in only _____ direction across a synapse.

one	3-201	Have you ever been examined by a physician and had your knee-jerk REFLEX tested? The physician taps your knee slightly below the kneecap--this is the stimulus; an impulse is transmitted; and you _____ by jerking your leg upward.
respond (or react)	3-202	The impulse has travelled -- in one direction -- from a receptor, your knee, first to the spine, and then to an effector organ, the muscle of your _____.
leg	3-203	A neuron which receives a stimulus is called a SENSORY neuron. When the doctor hit you, the stimulus was picked up by a sensory neuron in your _____.
knee	3-204	The sensory neuron routed the impulse to the spinal cord, which passed it on to a MOTOR neuron. The impulse then travelled via motor neurons to the _____ of your leg, and you kicked back.
muscle	3-205	You now have all the elements of the reflex pathway: Stimulus received by _____ neuron; impulse transmitted to spine; finally from spine to muscle via _____ neuron.
sensory motor	3-206	If your hand is being burned, you don't think about it -- you move to correct the situation. This is another example of an "arc" or _____ pathway.
reflex	3-207	When you are pricked with a pin, you may jump or yell. The stimulus, the pinprick, is picked up by a (1)_____ neuron and the (2) _____ arc is complete when a (3)_____ neuron makes muscles work.

(1) sensory (2) reflex (3) motor	3-208. If you respond to a pinprick by <u>both</u> jumping and yelling, this means that the axon of the sensory neuron transmitted the impulse to more than <u>one</u> _____ - neuron -- in other words, that more than one synapse was involved.
motor	3-209. The red circles indicate the 2 points of impulse transmission or _____.
synapses	3-210. A reflex arc is an established pattern of stimulus response. Sometimes a new stimulus is presented, and you are slow to respond; this is because there is no reflex arc established for the impulse, and it must be sent to the _____ for analysis.
brain	3-211. When someone tries to punch you in the eye, you automatically close your eyelid. This is an example of a _____ _____.
reflex arc (or reflex act)	3-212. An <u>impulse</u> can be compared to <u>electricity</u> that is produced in your automobile battery. The impulses that travel along a neuron are electrical in nature and are caused by the charged atoms or _____ inside and outside a neuron.
ions	3-213. Another electrical analogy can be made concerning the sheath that covers the axon. The sheath is similar in one way to the _____ that protects an electrical wire.

insulation	3-214. The myelin and neurolemmal sheaths are <u>unlike</u> electrical insulation because they permit ions to enter and leave the neuron, creating an _____ charge which allows an impulse to flow along a neural pathway.
electrical	3-215. Neural pathways are composed of tissues because the nerve cells which make up the pathways, like those of all tissues, have a similar _____ and _____ .
structure function	3-216. Do you recall, however, that sometimes <u>unlike</u> tissues combine? Each muscle cell is separate and is surrounded by a delicate connective tissue. The separately contained cells are a combination of two _____ tissues.
unlike	3-217. When unlike tissues combine, an <u>ORGAN</u> may be formed. For example, your heart is composed of muscle tissue, nervous tissue, and connective tissue. The heart is therefore an _____ .
organ	3-218. An organ is an organization of dissimilar _____ .
tissues	3-219. Your stomach is composed of connective, muscle, and nerve tissue. It is, like your heart, an _____ .
organ	3-220. Organs can combine together so that they make a functional unit. An example of this would be your digestive <u>SYSTEM</u>, which consists of your esophagus, stomach, intestines, and, some accessory or helping organs. (No response required. Move to frame 221.)

3-221.   The brain and spinal cord are both _____ of the nervous _____.

---

organs

system

3-222.   Chapter 4 will introduce you to a system composed of several organs, the <u>INTEGUMENTARY</u> (in-teg-u-MEN-ta-ri) <u>SYSTEM</u>. The skin is the integument of your body.

91

Skin and Skeleton
Part 1 -
The Integumentary System

4-1.   The word "integument" means "skin" or "covering". Your skin, nails, hair and certain glands form the INTEGUMENTARY (in-teg-u-MENT-a-ree) SYSTEM of your body, the subject of this unit.

(no answer required)

4-2.   Let's begin by reviewing some things you have already learned.

Cells that have a similar anatomy and physiology group together to form _tissues_ .

tissue

4-3.   An organization of dissimilar tissues is called a(n) _organ_ .

organ

4-4.   A group of organs that form a functional unit is a(n) _system_ .

system

4-5.   Look at your skin. Obviously the tissue you are looking at has one surface attached to deeper tissues and one surface exposed. Although you may see blood vessels <u>through</u> the tissue, the surface tissue itself is avascular. <u>Therefore</u> you know that the surface tissue of your skin is _epithelium_ (what kind of tissue?).

epithelium (epithelial tissue)	4-6. If you could peel back the epithelial tissue on the surface of your skin, you would see another layer of tissue which is also part of your skin. This second layer of tissue is highly vascular. It serves to support the surface tissue and to connect it to underlying tissue. The second layer of your skin, then, is composed of (epithelial/connective) _connective_ tissue.
connective	4-7. If you answered these five frames correctly, you are doing fine. If your answers were not correct, you might do well to pause for some review of Chapter 2 and 3 (No answer required. Go on to frame 8 or review, as your conscience dictates.)
	4-8. On the surface of your skin is a layer of _epithelial_ tissue. Beneath this surface is a layer of _connective_ tissue.
epithelial connective	4-9. You have learned that the skin is composed of two different types of tissue. An organization of dissimilar tissues is an organ. The skin, therefore, is an _organ_.
organ	4-10. The surface layer of the skin is called the _epider_. The inner layer of the skin is called the _dermis_. EPIDERMIS DERMIS SKIN
epidermis dermis	4-11. A note on word origins: The dermis is sometimes called the "true skin." The word dermis comes from the Greek word for skin, and the Greek prefix "epi" means "on" or "over." The epidermis is, literally, an "over-skin." You can easily remember that the epidermis consists of epithelial tissue, because both words begin with the same prefix, "epi". (no answer required)

	4-12. Another note on word origins:
	"Cuticle" (literally, "little skin") and "cutaneous" (pertaining to the skin) are both derived from the <u>Latin</u> word for "skin." The Latin prefix "sub" means "under." Thus we need not even tell you that the <u>SUBCUTANEOUS</u> (sub-kew-TA-ne-us) <u>TISSUE</u> is the layer of tissue <u>under</u> the skin.
under	4-13. Beneath the dermis of the skin is a fibro-fatty layer of tissue known as <u>subcutaneous tissue</u>.
subcutaneous

tissue | 4-14.   Label the three parts of this drawing:
 1. *epidermis*
 2. *dermis*
 3. *subcutaneous tissue*
 SKIN |
| 1. epidermis
2. dermis
3. subcutaneous
    tissue | 4-15.   The surface layer of the skin, the epidermis, consists of from two to four layers. On a thick-skinned surface, such as the palm of the hand or sole of the foot, the epidermis will have a maximum of (how many?)
 1. _4_ layers, while on even the thinnest-skinned surface, it will have a minimum of
 2. _2_ layers. |
| 1. four
2. two | 4-16.   The cells of the epidermis begin to develop or <u>germinate</u> in the bottom layer, called the <u>stratum germinativum</u>

 EPIDERMIS {
 STRATUM CORNEUM
 (STRA-tum COR-ne-um)

 STRATUM GERMINATIVUM
 (STRA-tum GER-mi-na-TE-vum) |

stratum germinativum	4-17. As new cells are produced, the older ones move toward the top layer of the epidermis, which is called the *stratum corneum*

EPIDERMIS { — STRATUM CORNEUM

— STRATUM GERMINATIVUM

stratum corneum	4-18. As the cells of the epidermis grow, they move toward the (inner/outer) *outer* layer.

outer	4-19. As the cells move from the stratum germinativum towards the stratum corneum, their protoplasm is gradually transformed into a horny substance, and the cells die. The stratum corneum then consists of a layer of (living/dead) *dead* cells.

dead	4-20. The dead cells are often described as "horny" (semi-opaque). Since the Latin word for "horn" is "cornu" (as in "unicorn"), this horny substance probably inspired the name for the layer of cells in which it is found, the *stratum corneum*.

stratum corneum	4-21. When epidermis consists of only two layers, those layers are the lower or germinating layer and the surface layer of dead cells. Label this drawing of two-layered epidermis.

1. *Stratum corneum*

2. *Stratum germinativum*

} EPIDERMIS

1. stratum corneum 2. stratum germinativum	4-22. The horny substance found in the cells of stratum corneum is called KERATIN (KER-a-ten). Keratin is a non-living substance which replaces the _Protopla_ of the cells as they move toward the surface of the skin.
protoplasm	4-23. The horny substance found in the cells of the stratum corneum is called _Keratin_.
keratin	4-24. Label this drawing of the skin: 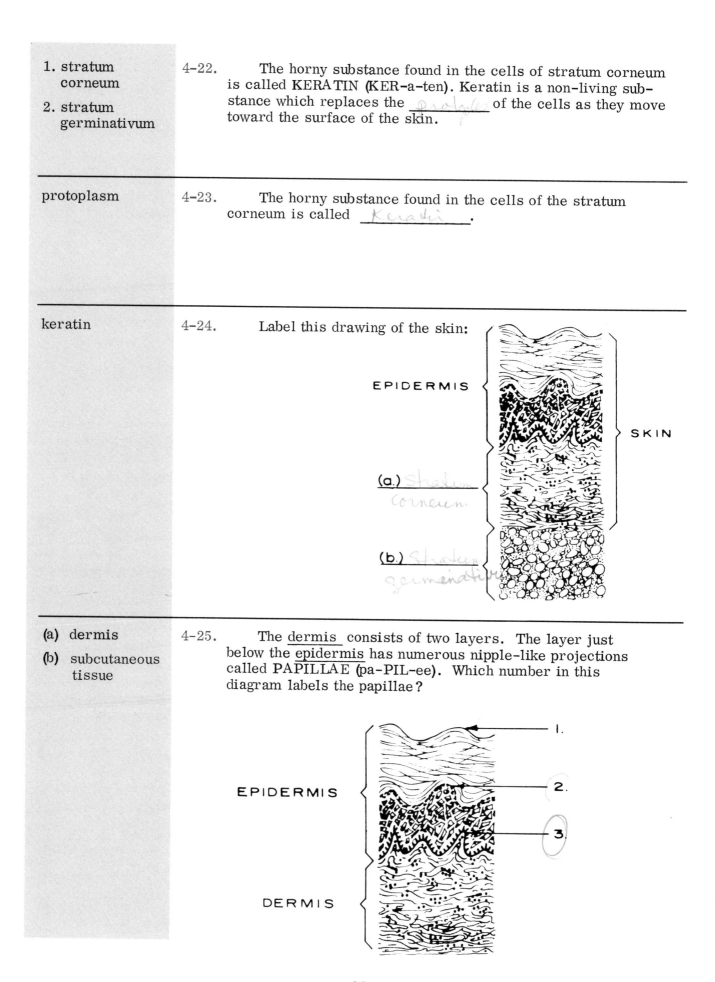 (a.) _Stratum Corneum_ (b.) _Stratum germinativum_
(a) dermis (b) subcutaneous tissue	4-25. The <u>dermis</u> consists of two layers. The layer just below the <u>epidermis</u> has numerous nipple-like projections called PAPILLAE (pa-PIL-ee). Which number in this diagram labels the papillae?

3.	4-26. The outer layer of the dermis is called the PAPILLARY (PAP-i-ler-ee) LAYER, and the inner layer, or basement membrane, is called the RETICULAR (re-TIC-u-ler) LAYER. Thus the layer of the dermis which touches the epidermis is the _papillary_ layer, and the layer which touches the subcutaneous tissue is the _reticular_ layer.
papillary reticular	4-27. The whorls on your fingers which create fingerprints result from the presence of projections called _papillae_ in the papillary layer of the dermis.
papillae	4-28. Voluntary movements of the face and neck are possible because there are skeletal muscles in the lower or _reticular_ layer of your dermis.
reticular	4-29. Label the following drawing: EPIDERMIS DERMIS 1. _stratum corneum_ 2. _stratum germinativum_ 3. _papillary layer_ 4. _reticular layer_
1. stratum corneum 2. stratum germinativum 3. papillary layer 4. reticular layer	4-30. Beneath the skin is a layer of tissue called subcutaneous tissue. Like the dermis, the subcutaneous tissue is a layer of _connective_ (what kind?) tissue.
connective	4-31. The subcutaneous tissue serves as a storage place for fat cells or adipose cells. Thus a fat or adipose person is one with an overabundance of (epidermis/subcutaneous tissue) _____.

subcutaneous tissue	4-32. The fat cells of the subcutaneous tissue are called _____ *adipose* _____ cells.
adipose	4-33. Until now, we have described the skin as an organ composed of layers of tissues. These tissues, however, are not simple structures laid one upon the next like a pile of blankets. As you will see, a network of glands, blood vessels, and nerves is woven through the tissues and functions with them. (no answer required)
	4-34. You will recall that a <u>vascular</u> tissue is a tissue which (contains/does not contain) _____ blood <u>vessels</u>.
contains	4-35. An <u>avascular</u> or <u>nonvascular</u> tissue is one which does not contain _____ *blood* _____ *vessels* .
blood vessels	4-36. You can see from the drawing below that the tissue of the epidermis is (vascular/avascular) _____ . EPIDERMIS DERMIS ————► BLOOD VESSELS ————►
avascular	4-37. The tissue of the dermis is (vascular/avascular) _____ .

98

vascular	4-38. Since the epidermis contains no blood vessels from which to obtain nourishment, it must receive its nutrients from the blood vessels closest to it; namely, the blood vessels in the upper or _papillary_ layer of the dermis.
papillary	4-39. Nutrients pass out of the blood vessels of the papillary layer and move up into the intercellular spaces of the _epidermis_.
epidermis	4-40. In addition to blood vessels, the dermis contains another system of vessels, known as <u>LYMPHATIC</u> (lim-FAT-tic) <u>VESSELS.</u> The lymphatic vessels remove tissue fluid from the tissue spaces. Thus the dermis receives its nutrients from the _blood_ vessels and excretes its excess tissue fluid into both the blood vessels and the _lymphatic_ vessels.
blood lymphatic	4-41. Both lymphatic vessels and blood vessels are found in great abundance in the (epidermis/<u>dermis</u>) _____.
dermis	4-42. Tissue fluid is carried away from the tissue spaces of the dermis by the blood vessels and by the _lymphatic vessels_.
lymphatic vessels	4-43. Differences in skin color exist between races, between individual members of the same race, and even between the various parts of one individual's body. Let us see what substances are responsible for these color differences. (no response required)
	4-44. A substance called MELANIN (MEL-a-nin) is responsible for brown color in skin. Freckles and brown moles occur at spots where the skin contains a high concentration of _melanin_.

melanin	4-45. The yellow color of skin is due to the presence of CAROTENE (KAR-o-ten). The skin of Negroes contains a large amount of _melanin_ while that of Asians contains much _carotene_. The skin of Caucasians contains varying amounts of both melanin and carotene.
melanin carotene	4-46. Although the words "keratin" and "carotene" sound similar, they mean quite different things. The yellow color of skin indicates the presence of _carotene_. The horny substance found in the cells of the stratum corneum is _keratin_.
carotene keratin	4-47. When a pin pricks your skin, or when something hot or cold touches your skin, you are aware of the sensation. Thus you can infer that your skin (contains/does not contain) _____ numerous nerves.
contains	4-48. Your skin serves as a sense organ because of _nerves_ located in the dermis.
nerves	4-49. The integumentary system contains numerous glands. Among these are the sweat glands, which produce _sweat_.
perspiration (sweat)	4-50. When the body is exposed to heat, it produces perspiration. The evaporation of the perspiration helps to regulate body temperature by making the skin (cooler/warmer) _____.

cooler	4-51 The sweat glands of the integumentary system help to regulate body _____temp_____ .

SWEAT GLAND

temperature	4-52 Both skin and sweat glands are members of the __integumentary__ system.
integumentary	4-53 CERUMINOUS (ser-OO-min us) GLANDS are modified sweat glands which produce CERUMEN (ser-OO-min) or ear wax. Ceruminous glands are located in the external __ear__ .
ear	4-54 Ear wax is the product of the __cerumen__ glands in the ear.
ceruminous	4-55 MAMMARY (MA-ma-re) GLANDS are located in the mammae (singular: mamma) or breasts. The main function of these specialized glands is the production of __milk__ .
milk	4-56 A nursing infant is nourished on the milk produced by the __mammary__ glands of its mother.

mammary	4-57	SEBACEOUS (se-BA-shus) GLANDS lubricate the skin and hair by producing an oil called sebum. If too little sebum is produced by the sebacious glands, a very (oily/dry) _____ skin may result.
dry	4-58	The function of the sebaceous glands is to __lubricate__ the skin and hair.
oil or lubricate	4-59	Certain glands help to control body temperature by secreting liquid from the skin. These are called __sweat__ glands.
sweat	4-60	Mammary glands produce 1.__milk__. Sebaceous glands produce 2.__sebum, oil__. Ceruminous glands produce 3.__wax__.
1. milk 2. oil 3. ear wax	4-61	So far, you have studied two components of the integumentary system. These two components are __skin__ and __glands__.
skin glands	4-62	Hair, like skin and glands, is a part of the __integumentary__ system.

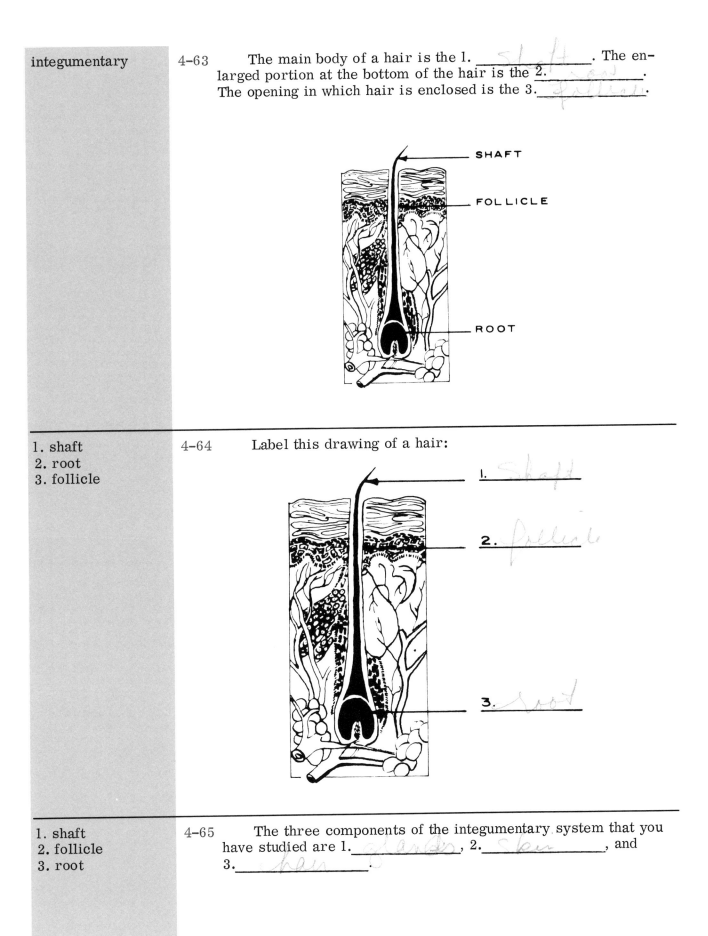

integumentary

4-63   The main body of a hair is the 1. _Shaft_ . The enlarged portion at the bottom of the hair is the 2. _root_ . The opening in which hair is enclosed is the 3. _follicle_ .

SHAFT

FOLLICLE

ROOT

1. shaft
2. root
3. follicle

4-64   Label this drawing of a hair:

1. _Shaft_

2. _follicle_

3. _root_

1. shaft
2. follicle
3. root

4-65   The three components of the integumentary system that you have studied are 1. _glands_, 2. _skin_, and 3. _hair_.

1. skin 2. glands 3. hair (any order)	4-66 Hair is a specialized form of skin; so, too, are fingernails and toenails. It is a common belief that nails are a form of bone; this belief is (correct/incorrect) _____.
incorrect	4-67 Both nails and hair are specialized forms of _skin_____.
skin	4-68 At the base of each nail is a small extension of the upper layer of epidermis. This "little skin", you know from experience, is called the _cuticle___.
cuticle	4-69 The prefix "epo" means "over" or "on". You can deduce, therefore, that the term EPONYCHIUM (ep-o-NIK-i-um) applies to (the skin under the nail/ the cuticle over the nail)_____.
the cuticle over the nail	4-70 The prefix "hypo" means "under". The term HYPONYCHIUM (hy-po-NIK-i-um) refers, therefore, to the skin (over/under) _under_____ the free edge of the nail.
under	4-71 _eponychium_ 1. _cuticle___ 2. _hyponychium_ Label the parts of this drawing of the fingernail as either eponychium or hyponychium.
1. eponychium 2. hyponychium	4-72 The skin beneath the free edge of the fingernail is called 1. _hyponychium_. The cuticle at the base of the nail is called the 2. _eponychium_.

1. hyponychium 2. eponychium	4-73	The cells of the fingernails contain keratin, a substance also present in the cells of the stratum _corneum_.
corneum	4-74	The presence of sulphur makes keratin hard and also hardens the tissue in which the keratin is found. Thus the keratin in the relatively soft epidermis is low in __sulphur__ while the keratin in the relatively hard nails is __high__ in sulphur.
sulphur high	4-75	The keratin in the nails contains a relatively (large/small) _____ amount of sulphur. The keratin in the epidermis contains a relatively (large/small) _____ amount of sulphur.
large small	4-76	The nails are part of the __integumentary__ system.

integumentary	4-77	You have learned the four components of the integumentary system. These components are:
		   1. _hair_ 3. _glands_
		   2. _nails_ 4. _skin_

1. skin 2. glands 3. hair 4. nails (any order)	4-78	At some time, you must accidentally have scraped off the outer layer of your skin. The exposed tissues then proved to be (more/less) _____ sensitive to stimuli than the skin itself.
more	4-79	If the tissues beneath your skin were not protected by the skin, you would be excessively sensitive to stimuli. One function of the skin is to __protect__ underlying tissues from injurious stimuli.

protect	4-80	The tissues beneath the skin are generally more susceptible to infection than is the skin. A covering of skin protects these underlying tissues not only from injurious stimuli but also from _infection_.
infection	4-81	If water were not held in the underlying tissues by the skin, the underlying tissues would become too dry. The skin prevents this drying of underlying tissues by covering the tissues and preventing excessive evaporation, or loss of _water_.
water	4-82	Two functions of the skin, then, are to _protect_ underlying tissues from injurious stimuli and to prevent the drying of underlying tissues by preventing excessive evaporation or loss of _water_.
protect water	4-83	In a hot environment the blood vessels of the skin dilate, or expand, so that more blood will reach the surface area of the skin and give off heat. As the heat is lost through the skin, the body temperature (increases/decreases) _____.
decreases	4-84	In a cold environment, the blood vessels constrict to prevent blood from reaching the surface and losing heat. Thus in both heat and cold, the blood vessels of the skin help to regulate the _____ of the body.
temperature	4-85	The nerve endings of the skin are sensitive to stimuli. Thus the skin, like the eye or ear, serves as a kind of _____ organ.
sense	4-86	Through the action of the sweat glands and through the contraction and dilation of blood vessels, the skin helps to regulate the _____ of the body.

temperature	4-87	Because it is better able to resist injurious stimuli than are the underlying tissues, the skin _____ the body.
protects	4-88	A short review will help you to remember what you have learned in this unit. The upper layer of the skin is called the _____, and the lower layer of the skin is called the _____.
epidermis dermis	4-89	Beneath the dermis of the skin is a layer of tissue called _____ _____.
subcutaneous tissue	4-90	Epidermis may consist of four layers, but it always consists of at least two layers. The surface layer of epidermis is called the _____ _____,and the lowest layer is called the _____ _____.
stratum corneum stratum germinativum	4-91	The upper layer of the dermis contains small projections which rise into the epidermis. These projections are called _____.
papillae	4-92	The upper layer of the dermis is called the _____ _____,and the lower layer is called the _____ _____.
papillary layer reticular layer	4-93	The epidermis consists of _____ tissue. The dermis consists of _____ tissue.

1. epithelial 2. connective	4-94	The skin is a(n) (tissue/organ/system) _____.
organ	4-95	The cells of the stratum corneum contain a horny substance called _____.
keratin	4-96	The fat cells of the subcutaneous tissue are called _____ _____.
adipose cells	4-97	The epidermis is (vascular/avascular) _____. The dermis is (vascular/avascular) _____.
avascular vascular	4-98	The epidermis receives its nutrients from the blood vessels of the _____ layer.
papillary	4-99	Tissue fluid is removed from the tissue spaces of the dermis by blood vessels and by _____ _____.
lymphatic vessels	4-100	The brown color of skin is caused by _____. The yellow color of skin is caused by _____.

melanin carotene	4-101	The glands which help to regulate body temperature by se-creting moisture are called the _____.
sweat glands	4-102	The ceruminous glands produce 1. _____. The sebaceous glands produce 2. _____. The mammary glands produce 3._____.
1. ear wax 2. oil 3. milk	4-103	The skin opening in which a shaft of hair is enclosed is called the _____ of the hair.
follicle	4-104	Hair and nails are both specialized forms of _____.
skin	4-105	The skin under the nail is called the _____. The cuticle over the nail is called the _____.
hyponychium eponychium	4-106	The technical term for the cuticle is the _____.
eponychium	4-107	The hard keratin of the nails contains (more/less) _____ sulphur than the soft keratin of the epidermis.

more	4-108 The components of the integumentary system are: 1. _____ (name them) 2. _____ 3. _____ 4. _____
1. skin 2. glands 3. nails 4. hair (any order)	4-109 Which of the following are functions of the integumentary system? 1. to protect underlying tissues 2. to store food and water 3. to serve as a sense organ 4. to act as a temperature regulator
all	4-110 You have finished the first part of this chapter. Take a break and relax! (no response required!)

Skin and Skeleton
Part 2 -
The Musculo- Skeletal System

4-111.   The subject of this section of the program will be the MUSCULO-SKELETAL (MUS-kew-lo-SKEL-e-tal) SYSTEM, which, as its name implies, is made up of the _muscles_ and the bones of the _skeleton_.

---

muscles
skeleton

4-112.   The various movements of which the body is capable result from the action of flexible muscles pulling against the hard, inflexible bones of the _skeleton_.

---

skeleton

4-113.   Muscles are attached to the bones of the skeleton by means of the fibrous cords or bands called tendons. The large muscle in the calf of the leg is attached to the "heel bone" by the Achilles _tendon_.

---

tendon

4-114.   Cartilage, you remember, is a somewhat flexible tissue which forms the connecting structures of the bony skeleton and often acts as a "shock absorber" between bones. Muscles are bound to bone by means of (tendons/cartilage) _tendons_. Between some bones are shock absorbers made of (tendons/ cartilage) _cartilage_.

---

tendons

cartilage

4-115.   Ligaments are the bands of tissue which bind the bones together. The bones are separated and connected by shock absorbers of _cartilage_ and are bound together by bands of _ligaments_

cartilage ligament (s)	4-116. The human skeleton contains <u>206</u> bones. Six of these bones are located in the ears, three in each ear. The remaining <u>_200_</u> (how many?) bones are divided between the AXIAL (AX-e-al) SKELETON and APPENDICULAR (ap-pen-DIK-u-lar) SKELETON.
200	4-117. The <u>appendicular</u> skeleton consists of all the bones of the <u>limbs</u> or <u>appendages</u>. The <u>axial</u> skeleton consists of all the remaining bones, except the six bones in the ears. The bones of the arm are part of the _appendicul_ skeleton. The bones of the head are part of the _axial_ skeleton.
appendicular axial	4-118. The spinal column is a part of the _axial_ skeleton. The bones of the foot are part of the _appendicul_ skeleton.
axial appendicular	4-119. The human body contains (2006/206) _206_ bones.
206	4-120. In addition to the 206 bones of the body, there are in most people a varying number of bones called <u>SESAMOID</u> (SES-a-moid) bones. Since the name "sesamoid" means "like a grain of sesame," you can infer that the sesamoid bones are quite (large/small) _small_.
small	4-121. Sesamoid bones may function as ball bearings to reduce friction in places where a muscle is joined to a bone. The sesamoids develop within the _tendons_ which join muscle to bone.
tendons	4-122. In addition to their ball-bearing function, the _sesamoid_ bones may also act as pulleys, affecting the direction of muscle pull.

sesamoid	4-123. Some sesamoids, such as those in the knee caps, are always present. In the joints of the hand and foot, sesamoids are usually present. In other areas of the body, sesamoids develop according to occupational demand. Thus, the number of sesamoids in the body (is/is not) ___is not___ the same for all humans.
is not	4-124. Bones may be classified in three ways, on the basis of: (1) shape, (2) structure, or (3) embryonic development. When we speak of bones as flat, long, irregular or short, we are obviously using a classification based on ___shape___.
shape	4-125. A section of a bone examined under a microscope shows its structure. Structurally, the illustrated bone has an outer layer of ___compact bone___ and an inner layer of ___spongy bone___.
	SPONGY BONE COMPACT BONE
compact bone spongy bone	4-126. Compact bone is dense and hard. Spongy bone has many small cavities filled with marrow. The classification of bones as compact or spongy is based upon their (shape/structure) ___structure___.
structure	4-127. Spongy bone is also called CANCELLOUS (CAN - sel - us) bone. The (inner/outer) ___inner___ layer of the illustrated bone is cancellous.
	SPONGY BONE COMPACT BONE

inner	4-128. All bones contain both compact and spongy tissue, but one predominates, and the bone is classified on the basis of the predominating type of tissue. Thus the classification of a bone as "spongy" indicates that the bone: (a) contains no compact tissue. (b) contains more spongy tissue than compact tissue.
(b) contains more spongy tissue than compact tissue	4-129. Following conception, bone, like all tissues, originates as soft tissue. Actual bone formation in the embryo occurs during the last two months of pregnancy and continues until maturity (18-21 years). The bones of the embryo develop in one of two ways, which we shall now discuss. (no answer required)
	4-130. The drawings illustrate three early stages of the growth of a long bone. The long bone originates as a cartilage _model_. Gradually the cartilage becomes _ossified_. CARTILAGE MODEL CARTILAGE
model ossified	4-131. The bone illustrated on the last item is, on the basis of its development, classified as CARTILAGINOUS (car-ti-LAH-ji-nus). A cartilaginous bone, as its name implies, is a bone which develops from a _cartilage_ _model_.
cartilage model	4-132. The bones of the face and cranium develop by another method as illustrated here. First, bony spicules arise within a connective tissue (1) _membrane_. Then the (2) _spicules_ increase in number. Ultimately, they grow together or (3) _coalesce_. CONNECTIVE TISSUE MEMBRANE BONY SPICULES BONE (Spicules have ossified)

1. membrane 2. bony spicules 3. coalesce	4-133. The last item illustrated the development of <u>MEMBRANOUS</u> (MEM – bra – nus) bone. Membranous bone develops as ___*bony*___ spicules form and ultimately coalesce within a connective tissue ___*membrane*___
bony membrane	4-134. Bone which develops from a cartilage model is called ___*cartilaginous*___ bone. Bone which develops from bony spicules within a connective tissue membrane is called ___*membranous*___ bone.
cartilaginous membranous	4-135. The classification of bones as flat, irregular, long, or short is based upon the ___*shape*___ of the bones. The classifi- cation of bones as spongy (cancellous) or as compact is based upon the ___*structure*___ of the bones.
shape structure	4-136. The classification of bones as cartilaginous or as membranous is based upon the way in which the bones ___*develop*___.
develop (form, etc.)	4-137. Under a microscope, a longitudinal section of a typical bone looks like this: VOLKMANN'S CANAL HAVERSIAN CANALS Bone tissue is not perfectly solid but is traversed by a system of ___*canals*___.
canals	4-138. Blood vessels, lymph vessels and nerves enter a bone by means of its numerous ___*canals*___.

canals

4-139. A cross section of bone reveals an inner cavity called a ___marrow cavity___ and an outer sheath of tissue called the ___periosteum___

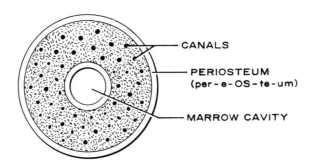

CANALS

PERIOSTEUM
(per - e - OS - te - um)

MARROW CAVITY

marrow cavity

periosteum

4-140. The connective tissue sheath which covers a bone is the periosteum. The connective tissue which lines the marrow cavity is the ___endosteum___

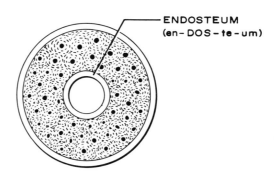

ENDOSTEUM
(en - DOS - te - um)

endosteum

4-141. Label the parts of this cross section of bone.

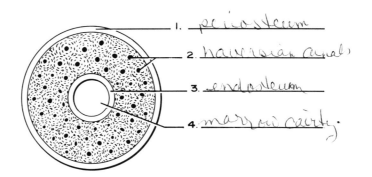

1. ___periosteum___

2. ___haversian canals___

3. ___endosteum___

4. ___marrow cavity___

1. periosteum 2. canals 3. endosteum 4. marrow cavity	4-142.　　The <u>articular</u> end of a bone is the end which makes a joint with another bone. The periosteum covers all of a bone except its _articular ends_ . ARTICULAR ENDS OF BONES PERIOSTEUM
articular end (s)	4-143.　　The parts of a long bone (such as the bones of the arm or leg) are classified as the (1) <u>DIAPHYSIS</u> (di - A F - i - sis), the shaft or main body of the bone, and (2) the <u>EPIPHYSIS</u> (ep - IF - i - sis), the end of the bone. Label the parts of this long bone. 1. _epiphysis_ 2. _diaphysis_
1. epiphysis 2. diaphysis	4-144.　　The prefix "dia" means "dividing in two." In a long bone, the _diaphysis_ is the long shaft which divides or separates the two epiphyses (plural of epiphysis).
diaphysis	4-145.　　The periosteum covers all of a bone except its articular end. In a long bone, the periosteum covers all of the (epiphysis/diaphysis) _diaphysis_ but may not cover all of the (epiphysis/diaphysis) _epiphysis_ .
diaphysis epiphysis	4-146.　　Bones receive their nutrients from the blood vessels, which traverse the bone canals. These blood vessels enter the bone through the outer sheath of the bone, which is called the _periosteum_ .

periosteum	4-147. If the periosteum is damaged, the blood vessels are disrupted, and the bone loses its blood supply. Obviously, the integrity of the periosteum is (essential to/not essential to) _essential to_ the health of the underlying bone.
essential to	4-148. The largest bones of the limbs are classified, on the basis of shape, as <u>long bones.</u> In the center of a long bone is a marrow cavity, which is so named because it contains bone _marrow_
marrow	4-149 <u>Yellow</u> bone marrow, composed chiefly of fat cells, is found in the marrow cavities of _long_ (what shape?) bones, like the large bones of the limbs.
long	4-150. In the marrow cavities of long bones, there is _yellow_ (what color?) marrow which consists chiefly of _fat_ cells.
yellow fat	4-151. The <u>epiphyses</u> of long bones are composed of spongy bone filled with <u>red</u> marrow. The marrow in the central cavity of a long bone is _yellow_, while the marrow in the spongy ends of the bone is _red_.
yellow red	4-152. Red marrow, found in the spongy bone of long bones and also in some flat bones, produces blood cells. One function of the bony skeleton, then, is to produce _blood cells_.
blood (cells)	4-153. In addition to its function of housing marrow, which produces blood cells, the skeleton forms the framework of the body and provides attachments for skeletal _muscles_.

muscles	4-154. The bones of the cranium enclose and protect the brain. The bones of the thorax enclose and protect the heart, lungs, and other vital organs. Therefore, another function of the skeleton is _protection_ .(Use your own words.)
to enclose and protect the vital organs. (or any similar statement)	4-155. A machine whose moving parts come in contact with each other must have lubrication at the points of contact. Bones come together at joints called <u>ARTICULATIONS</u> and, where movement is involved, also require _lubrication_ as does a machine.

lubrication	4-156. Between the <u>mobile</u> bones of the limbs,there are articulations. There are also articulations between the <u>immobile</u> bones of the skull. At which articulations do you think there is a greater need for lubrication?
	
	    (a) those of the skull
	    (b) those of the limbs

(b) those of the limbs	4-157. Articulations such as those of the limbs and spine are lubricated by <u>SYNOVIAL</u> (si – NO – vi – al) FLUID,which is contained in a cavity between bones. Taking their name from the fluid which lubricates them, these articulations are called _synovial_ joints.
synovial	4-158. The articulations of the limbs, fingers, and bones of the spine are lubricated by _synovial fluid_ .
synovial fluid	4-159. Before you go on, review what you have learned so far. The human skeleton contains (2006/206) bones _206_ .
206	4-160. The bones of the limbs are part of the _appendicular_ skeleton. All the remaining bones in the body except the six bones in the ears are part of the _axial_ skeleton.

119

appendicular
axial

If you made an
error, review
frames 4-116
to 4-118.

4-161.   In most people there are a varying number of small
bones which may function as ball bearings in places where
muscle is joined to bone or as pulleys affecting the direction
of muscle pull. These small bones are called _sesamoid_
bones.

---

sesamoid bones

4-162.   The first column lists some terms used to classify
bones. The second column lists the bases of bone classifi-
cation. Match the items in the two columns.

(1) spongy and compact
     bones
(2) flat, round, long, short
     and irregular bones
(3) membranous and
     cartilaginous bones

(a) embryonic development

(b) shape

(c) structure

If you made an
error, review
frames 4-120
to 4-123.

---

1. c
2. b
3. a

4-163.   Label the parts in the drawing of the cross section
of a bone.

1. _periosteum_
2. _canals_
3. _endosteum_
4. _marrow cavity_

If you made an
error, review
frames 4-124
to 4-136.

---

1. periosteum
2. canals
3. endosteum
4. marrow cavity

4-164.   Label the two parts of this drawing of a long bone as
either diaphysis or epiphysis.

1. _epiphysis_
2. _diaphysis_

If you made an
error, review
frames 4-137
to 4-141.

	4-165. Articulations, such as those of the limbs and spine, are lubricated by a fluid called _synovial fluid_.
	4-166. Now that you have examined the structure, functions, and classification of bones, consider how these bones are arranged to form the skeleton, which supports the body. (no answer required)
	4-167. The skull consists of 22 bones; eight of these bones form the CRANIUM (KRA - ni - um), which houses the brain, and the remaining fourteen form the face. The terms skull and cranium are not synonyms. Both the _cranium_ and the face are parts of the _skull_.
cranium skull	4-168. The bones of the cranium form the CRANIAL VAULT. The cranial vault houses the _brain_.
brain	4-169. Except for the mandible which has a ball and socket articulation, bones of the skull are joined together by serrated joints called _suture_. CRANIUM — SUTURES NASAL CAVITY MANDIBLE
sutures	4-170. The bones of the skull are immovable except for the "jaw-bone," called the _mandible_, which is movable.

mandible

4-171.   Below the skull is the spine or _vertebral column_.

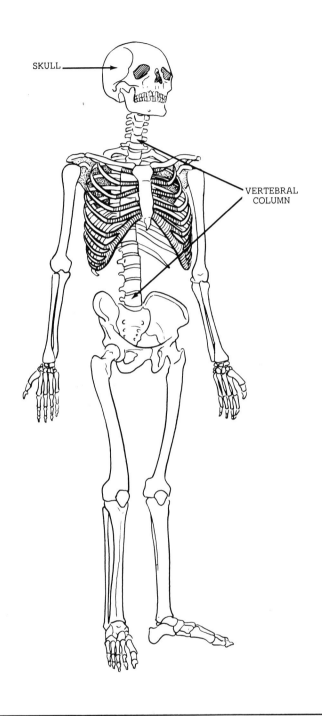

SKULL

VERTEBRAL
COLUMN

**vertebral
column**

4-172.   The vertebral column of an adult consists, usually, of 33
separate bones called <u>VERTEBRAE</u> (VER-te-bray), although
the number may vary by one in some individuals. The articu-
lations of these vertebrae make the spine (flexible/inflexible)
flexible.

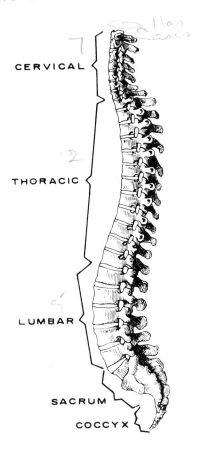

---

flexible

4-173.   Look at the drawing of the vertebral column, and use it
to answer this and the following items.

The vertebral column is divided into five regions. The region
just below the skull derives its name from the Latin word
"cervix" meaning "neck " and is called the _cervical_ region.

---

cervical

4-174.   The THORAX (THOR-ax) is the chest cavity. That section
of the vertebral column in the region of the chest is called the
thoracic region.

---

thoracic

4-175.   Just below the skull, at the neck, is the _cervical_
region of the vertebral column. Below that, in the chest
area, is the _thoracic_ region of the vertebral column.

123

cervical thoracic	4-176. Below the thoracic region of the vertebral column is the <u>Lumbar</u> region, its name meaning "loin."
lumbar	4-177. Below the lumbar region, the vertebrae have fused to form a single spade-shaped bone about which there arose a curious superstition. The bone is called the <u>SACRUM,</u> which means "sacred." It was so named because people once believed that the sacrum alone did not disintegrate after death and served as the basis for the resurrected body. (no answer required)
	4-178. Below the lumbar region, the vertebrae have fused to form a single bone, called the <u>sacrum</u>.
sacrum	4-179. At the bottom of the vertebral column, more fused vertebrae form the <u>COCCYX</u> (COCK-six). Coccyx means "cuckoo." The coccyx is so named because it was thought to resemble the beak of a cuckoo. (no answer required)
	4-180. Below the sacrum is the <u>coccyx</u>.

coccyx	4-181. It's quite easy to remember the names of the regions of the spine if you know what their names mean (and remember that the "sacred bone" is above the "cuckoo's beak"). Match the names on the left with their meanings on the right.
	
	     (1) sacrum a. chest
	
	     (2) coccyx b. neck
	
	     (3) cervical c. sacred
	
	     (4) lumbar d. loin
	
	     (5) thoracic e. cuckoo

1. sacrum – c.
   sacred

2. coccyx – e.
   cuckoo

3. cervical – b.
   neck

4. lumbar – d.
   loin

5. thoracic – a.
   chest

4-182.　Turn away from the illustration of the spinal column and do not refer to it as you answer this item.

Label this drawing of the regions of the spine:

1. Cervical

2. thoracic

3. lumbar

4. Sacrum

5. coccyx

---

1. cervical region
2. thoracic region
3. lumbar region
4. sacrum
5. coccyx

4-183.　The cervical, thoracic, and lumbar vertebrae are separate, movable bones, and so are called TRUE VERTEBRAE. The vertebrae of the sacrum and coccyx have fused together and so are called FALSE VERTEBRAE. The upper three regions of the spine are composed of (true/false) ____true____ vertebrae.

---

true

4-184　The true vertebrae make up the (1) ___cervical___, (2) ___thoracic___ and (3) ___lumbar___ regions of the vertebral column.

---

1. cervical
2. thoracic
3. lumbar
(any order)

4-185.　The term "false vertebrae" refers to the fused vertebrae of the ___sacrum___ and ___coccyx___.

125

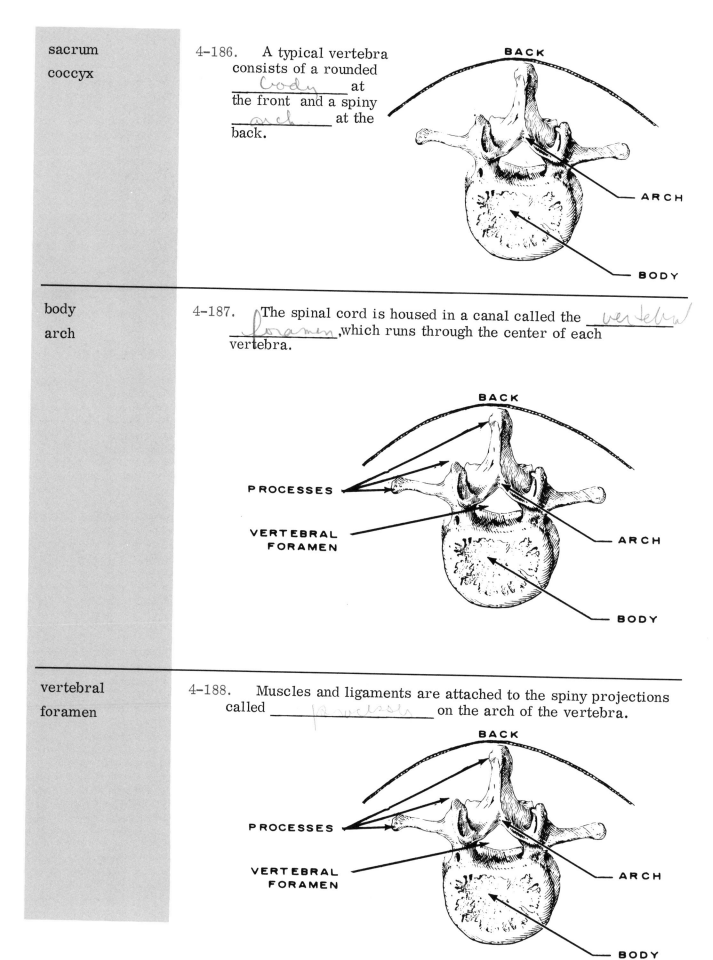

sacrum coccyx	4-186. A typical vertebra consists of a rounded _body_ at the front and a spiny _arch_ at the back.

BACK

ARCH

BODY

body arch	4-187. The spinal cord is housed in a canal called the _vertebral foramen_ ,which runs through the center of each vertebra.

BACK

PROCESSES

VERTEBRAL FORAMEN

ARCH

BODY

vertebral foramen	4-188. Muscles and ligaments are attached to the spiny projections called _processes_ on the arch of the vertebra.

BACK

PROCESSES

VERTEBRAL FORAMEN

ARCH

BODY

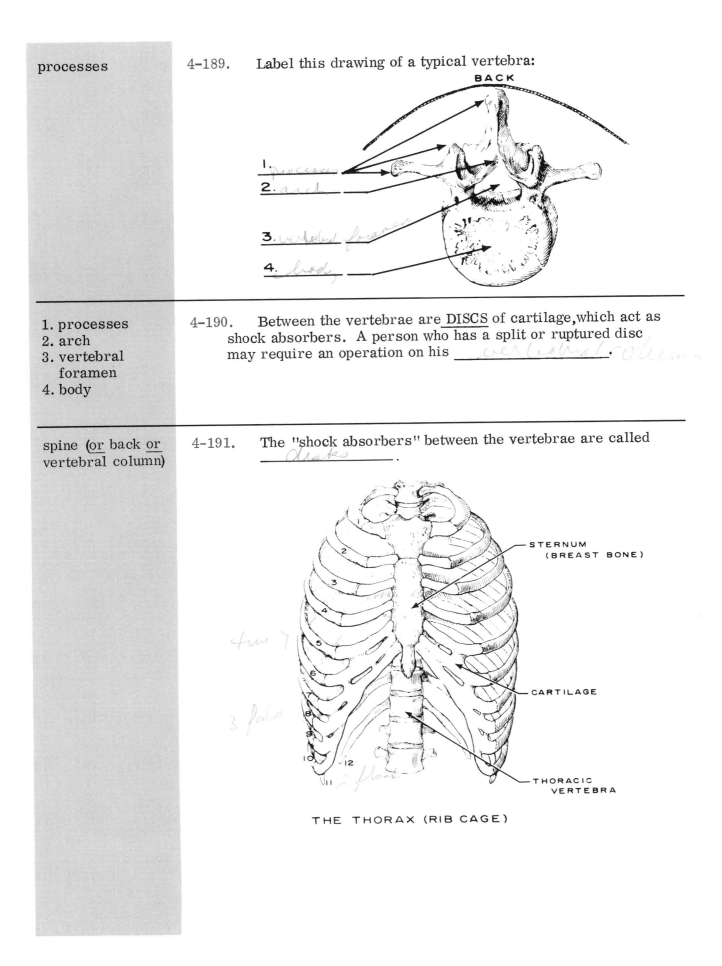

processes

4-189. Label this drawing of a typical vertebra:

BACK

1. _processes_
2. _arch_
3. _vertebral foramen_
4. _body_

1. processes
2. arch
3. vertebral
   foramen
4. body

4-190. Between the vertebrae are DISCS of cartilage, which act as shock absorbers. A person who has a split or ruptured disc may require an operation on his ___vertebral column___.

spine (<u>or</u> back <u>or</u> vertebral column)

4-191. The "shock absorbers" between the vertebrae are called ___discs___.

STERNUM
(BREAST BONE)

CARTILAGE

THORACIC
VERTEBRA

THE THORAX (RIB CAGE)

discs

4-192   Refer to the drawing of the thorax (on the preceding page),
and use it to answer this and the following items.

The ribs articulate with the ___*thoracic*___
vertebrae to form the thorax.

---

thoracic

4-193.   Those ribs which articulate with the STERNUM (breast
bone) as well as with the thoracic vertebrae are called TRUE
RIBS. There are ___7___ (how many?) true ribs.

---

seven

4-194.   The lowest five ribs are called FALSE RIBS because they
do not articulate with the ___*sternum*___.

---

sternum

4-195.   Because they articulate with the cartilage of the seventh
rib, the ___8___ , ___9___ , and ___10___ (what numbers?) ribs are
called VERTEBROCHONDRAL (VER-te-bro-KON-dral).

8th, 9th, 10th

4-196.   The front ends of the FLOATING RIBS are not articulated
(connected). The lowest ___2___ (how many?) ribs are
floating ribs.

---

two

4-197.   In the following ten frames you will see how the remaining
bones of the body combine to form the skeleton. The next
ten frames are intended to give you a general concept of
skeletal structure; you will not be expected to remember the
names of all the bones mentioned in these frames.

(no answer required)

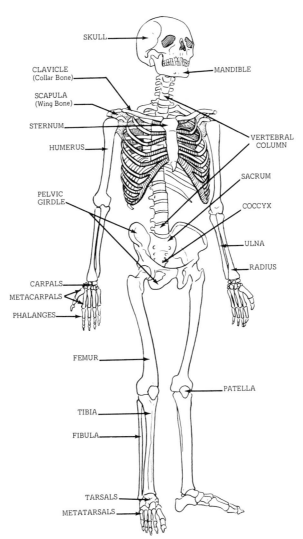

SKULL

MANDIBLE

CLAVICLE
(Collar Bone)

SCAPULA
(Wing Bone)

STERNUM

HUMERUS

VERTEBRAL
COLUMN

SACRUM

PELVIC
GIRDLE

COCCYX

ULNA

RADIUS

CARPALS
METACARPALS

PHALANGES

FEMUR

PATELLA

TIBIA

FIBULA

TARSALS
METATARSALS

SKELETON

4-198   Refer to the drawing of the skeleton and use it to answer this and the following items.

Refer to the drawing of the skeleton and use it to answer this and the following items.

The coalition of several bones, including the sacrum and coccyx, forms the _pelvic_ _girdle_.

pelvic girdle

4-199.   The bones of the pelvic girdle articulate with the _vertebral_ _column_ above and with the thigh bones or _femur_ below.

129

vertebral column femurs	4-200. The <u>hip joint</u> is formed by the articulation of the bones of ___vertebral___ ___column___ with the ___femur___ below.
pelvic girdle femur	4-201. Below the knee, the main bone of the leg is the (1) ___tibia___, which articulates with the (2) ___femur___ above.
1. tibia 2. femur	4-202. In addition to the tibia, the lower leg contains a smaller bone, the ___fibula___, which articulates with the tibia.
fibula	4-203. The lower extremity of the skeleton is completed by the ankle bones or (1) ___tarsals___, the foot bones or (2) ___metatarsals___ and the toe bones or (3) ___phalanges___.
1. tarsals 2. metatarsals 3. phalanges	4-204. In the upper part of the body, the "collar bone" or ___clavicle___ articulates with the "wing bone" or ___scapula___.
clavicle scapula	4-205. The bony structure of the arm is somewhat similar to that of the leg. The upper arm contains a single large bone called the ___humerus___
humerus	4-206. The humerus articulates with two bones, the ___ulna___ and the ___radius___ in the lower arm.

radius ulna (either order)	4-207.　The radius and ulna articulate with the bones of the wrist, which are called (1) _carpals_ . Below the carpals are the bones of the hand, called (2) _metacarpals_ and the bones of the fingers, called (3) _phalanges_
1. carpals 2. metacarpals 3. phalanges	4-208.　Consider the muscles which combine with the skeleton to form the musculo-skeletal system. In previous units, you learned that the body contains three kinds of muscle: <u>smooth</u> muscle, which lines the internal organs; <u>cardiac</u> muscle which is found in the heart; and <u>skeletal</u> muscle, which allows the body to _move_ .
move	4-209.　The musculo-skeletal system functions to support the body and to allow it to move. The principal components of the musculo-skeletal system are bones and _skeletal_ (what kind?) muscles.
skeletal	4-210.　Skeletal muscles contain bundles of fibers held together by connective tissue and abundantly supplied with blood vessels. Skeletal muscles, in other words, are (vascular/avascular) _vascular_ .
vascular	4-211.　The numerous nerves present in skeletal muscle permit voluntary movement of the muscles. A muscle may be unable to move if the _nerves_ in that muscle are damaged.
nerves	4-212.　Muscles may be attached <u>directly</u> to a bone, or else they may be attached indirectly, by means of a _tendon_ like the Achilles tendon in the lower leg.
tendon	4-213.　Typically, a muscle is attached at two points: an <u>ORIGIN</u>, which acts as a fixed point or anchor, and an <u>INSERTION</u>, which is the movable end of the muscle. When a muscle contracts, the movable end or _insertion_ is brought closer to the anchor or _origin_ .

131

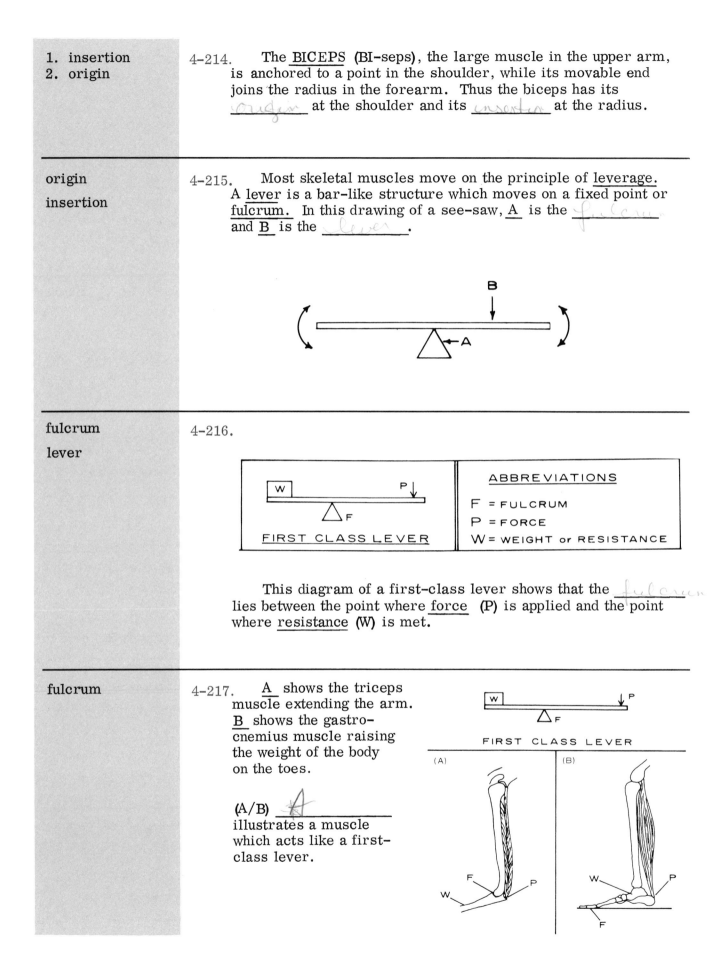

1. insertion 2. origin	4-214. The <u>BICEPS</u> (BI-seps), the large muscle in the upper arm, is anchored to a point in the shoulder, while its movable end joins the radius in the forearm. Thus the biceps has its <u>*origin*</u> at the shoulder and its <u>*insertion*</u> at the radius.
origin insertion	4-215. Most skeletal muscles move on the principle of <u>leverage</u>. A <u>lever</u> is a bar-like structure which moves on a fixed point or <u>fulcrum.</u> In this drawing of a see-saw, <u>A</u> is the <u>*fulcrum*</u> and <u>B</u> is the <u>*lever*</u> .
fulcrum lever	4-216. This diagram of a first-class lever shows that the <u>*fulcrum*</u> lies between the point where <u>force</u> (P) is applied and the point where <u>resistance</u> (W) is met.
fulcrum	4-217. <u>A</u> shows the triceps muscle extending the arm. <u>B</u> shows the gastro-cnemius muscle raising the weight of the body on the toes. (A/B) *A* illustrates a muscle which acts like a first-class lever.

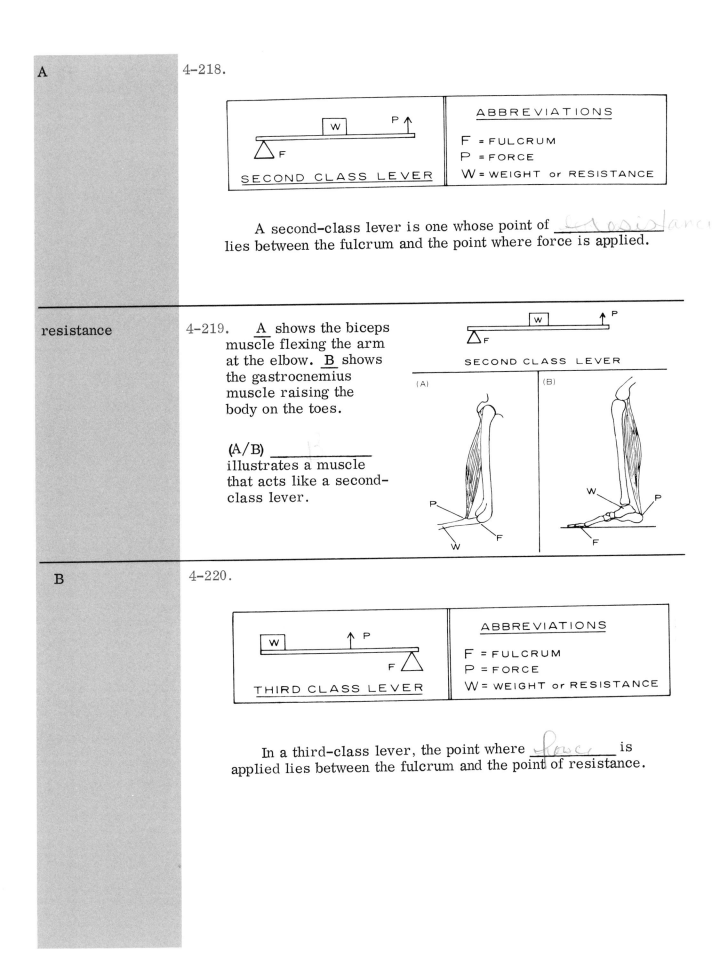

A

4-218.

ABBREVIATIONS

F = FULCRUM
P = FORCE
W = WEIGHT or RESISTANCE

SECOND CLASS LEVER

A second-class lever is one whose point of _____resistance_____ lies between the fulcrum and the point where force is applied.

---

resistance

4-219.   **A** shows the biceps muscle flexing the arm at the elbow. **B** shows the gastrocnemius muscle raising the body on the toes.

(A/B) _____ illustrates a muscle that acts like a second-class lever.

SECOND CLASS LEVER

(A)   (B)

B

4-220.

ABBREVIATIONS

F = FULCRUM
P = FORCE
W = WEIGHT or RESISTANCE

THIRD CLASS LEVER

In a third-class lever, the point where _____force_____ is applied lies between the fulcrum and the point of resistance.

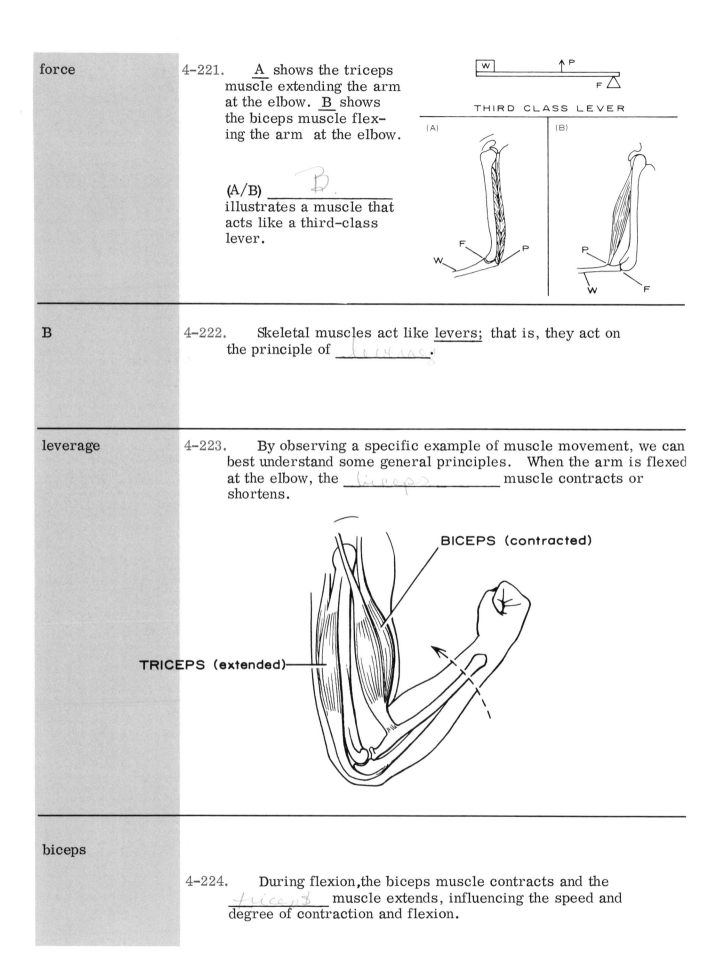

force

4-221.　 <u>A</u> shows the triceps muscle extending the arm at the elbow. <u>B</u> shows the biceps muscle flexing the arm at the elbow.

(A/B) _____B_____ illustrates a muscle that acts like a third-class lever.

THIRD CLASS LEVER

(A)

(B)

B

4-222.　 Skeletal muscles act like <u>levers</u>; that is, they act on the principle of _____leverage_____.

---

leverage

4-223.　 By observing a specific example of muscle movement, we can best understand some general principles. When the arm is flexed at the elbow, the _____biceps_____ muscle contracts or shortens.

BICEPS (contracted)

TRICEPS (extended)

---

biceps

4-224.　 During flexion, the biceps muscle contracts and the _____triceps_____ muscle extends, influencing the speed and degree of contraction and flexion.

triceps	4-225. Flexion of the arm results <u>chiefly</u> from the contraction of the biceps muscle, but the speed and degree of flexion are influenced by the triceps muscle and by other muscles in the arm. Thus flexion results from the action of (one muscle/a group of muscles) _a group of muscles._
a group of muscles	4-226. The example of muscle movement in flexion of the arm illustrates a general principle which is true of all muscles. Body movement results <u>not</u> from the action of an individual muscle, but from the action of a ___group___ of muscles.
group	4-227. Although a group of muscles acts together to move a part of the body, one muscle in the group usually bears the principal responsibility for the action. Flexion of the arm results principally from the contraction of the ___biceps___.
biceps	4-228. The muscle which bears the principal responsibility for the movement of a group of muscles is called the <u>PRIME MOVER</u> of the group. When the arm is flexed, the ___biceps___ is the prime mover.
biceps	4-229. <u>Extension</u> of the arm results principally from the action of the triceps. When the arm is extended, the triceps is the ___prime mover___.
prime mover	4-230. The ANTAGONIST of a muscle is another muscle which acts with an opposite effect. In flexion, the antagonist of the biceps muscle is the ___triceps___.
triceps	4-231. When the movement desired is extension of the arm, the triceps is the prime mover and the biceps is the ___antagonist___.

135

antagonist	4-232.　When the arm is flexed, certain groups of muscles called FIXATION MUSCLES hold the shoulder and other body parts steady to facilitate the action. Whenever any part of the body is moved, other parts are held steady by the action of _____ muscles.
fixation	4-233.　When someone lifts a heavy object, his legs, back, neck, etc. are held steady by _____.
fixation muscles	4-234.　When the fingers are being flexed by the flexor muscles in the forearm, certain muscles called SYNERGISTS (si-NER-jistz) prevent the unnecessary action of bending at the wrist. Synergists function by preventing _____ muscle action.
unnecessary	4-235.　The muscles which hold or fix body parts steady are called _____ muscles. The muscles which prevent unnecessary movements are called _____.
fixation synergists	4-236.　EMERGENCY MUSCLES are muscles which act to produce extra force or speed in a motion. When a hot object touches the sole, flexion of the foot is aided by leg and even thigh muscles acting as _____ muscles.
emergency	4-237.　The DELTOID (DEL-toid) muscle caps the shoulder like an epaulet. The fibers at the front of the deltoid flex the arm, while the fibers at the rear of the deltoid extend the arm. Different parts of the same muscle (can/cannot) _____ have opposite effects.
can	4-238.　The RECTUS MUSCLES, which extend from the sternum to the pelvic girdle, can flex the body. The ABDOMINAL MUSCLES can produce the same effect, namely flexion of the body. This example illustrates that two different muscles can have the _____ effect.

same	4-239. Unless a muscle has both (1) an intact nerve supply and (2) exercise, the muscle will waste or <u>ATROPHY</u> (A-tro-fe). Prolonged fixation of a broken bone in a plaster cast results in DISUSE ATROPHY, which is due to lack of __exercise__.
exercise	4-240. In addition to exercise, a muscle must have an intact nerve supply, or __disuse atrophy__ will result.
atrophy	4-241. Muscles are sheathed in a dense fibrous tissue called <u>DEEP FASCIA</u> (FA-sha). Deep fascia encloses blood and lymphatic vessels and serves to separate the individual __muscles__ which it sheathes.
muscles	4-242. Another name for the subcutaneous tissue beneath the skin is SUPERFICIAL FASCIA. The muscles are invested in sheaths of __deep fascia__. The layer of subcutaneous tissue beneath the dermis is called the __superficial fascia__.
deep fascia superficial fascia	4-243. <u>SUBSEROUS</u> (sub-SER-us) FASCIA lines the thoracic cavity. The lining of the abdominal cavity, like that of the thoracic cavity, is also composed of __subserous__ fascia.
subserous	4-244. Match the items in these two columns: 1. deep fascia (a) subcutaneous tissue 2. subserous fascia (b) forms the investing sheaths of muscles 3. superficial fascia (c) lines the thoracic and abdominal cavities
1. b 2. c 3. a	4-245 You have reached the end of Chapter 4. Take a break before you begin the next Chapter.

CHAPTER 5

The Respiratory System

5-1   In order to carry on its life functions, the human body must constantly receive oxygen from the air and must, in turn, eliminate carbon dioxide. RESPIRATION is the process through which the body accomplishes this intake of _____ and elimination of _____.

oxygen
carbon dioxide

5-2   In this unit you will learn about the RESPIRATORY SYSTEM, which is composed of all the organs involved in the vital process of _____.

respiration

5-3   The terms INSPIRATION and EXPIRATION refer to breathing in and to breathing out. Ordinary unstrained respiration is called QUIET RESPIRATION and consists of two phases, QUIET _____ in which air is inhaled and QUIET _____ in which air is exhaled.

inspiration
expiration

5-4   It is, of course, possible to breathe through one's mouth. Usually, however, human beings breathe through their _____.

noses (nostrils)

5-5   The nostrils, which are lined with short, stout hairs, are the outermost limits of the respiratory system. As air is inhaled through the _____, it is cleaned and filtered by the lining of _____.

nostrils hair (s)	5-6	The dilated part of the nostrils, just inside its opening, is called the VESTIBULE (VES-ti-bewl). The lining of hair is found only in the portion of the nostril just inside its opening, that is, in the _____.
vestibule	5-7	Above the vestibule, the nostrils are lined with a layer of mucous membrane from which grow hair-like projections of microscopic size called CILIA (SIL-ee-a). The cilia, like the hairs in the vestibule, serve to _____ the air which passes through them.
clean (or filter, etc.)	5-8	Large particles of matter are removed from the air as it passes the hairs in the vestibule. Smaller particles pass into the upper nostril where they become entrapped in _____ secreted by the mucous membrane lining.
mucus	5-9	Some of the particles entrapped in mucus move down and out of the vestibule; the remainder pass along a channel leading to the mouth. The mucus is moved toward the mouth by the beating of the hair-like _____.
cilia	5-10	As much as a quart of mucus may be secreted by the nasal membranes in one day. The fluid passes into the mouth where it can be expectorated or _____ out.
spit	5-11	Before air enters the lungs, it must be warmed and moistened, as well as cleaned. The mucous membrane in the nostrils helps not only to clean air but to _____ it and _____ it as well.
warm moisten (either order)	5-12	To help you form your own mental picture of the Respiratory System, look at the diagram of the Respiratory System. (Page 140) (No response required).

139

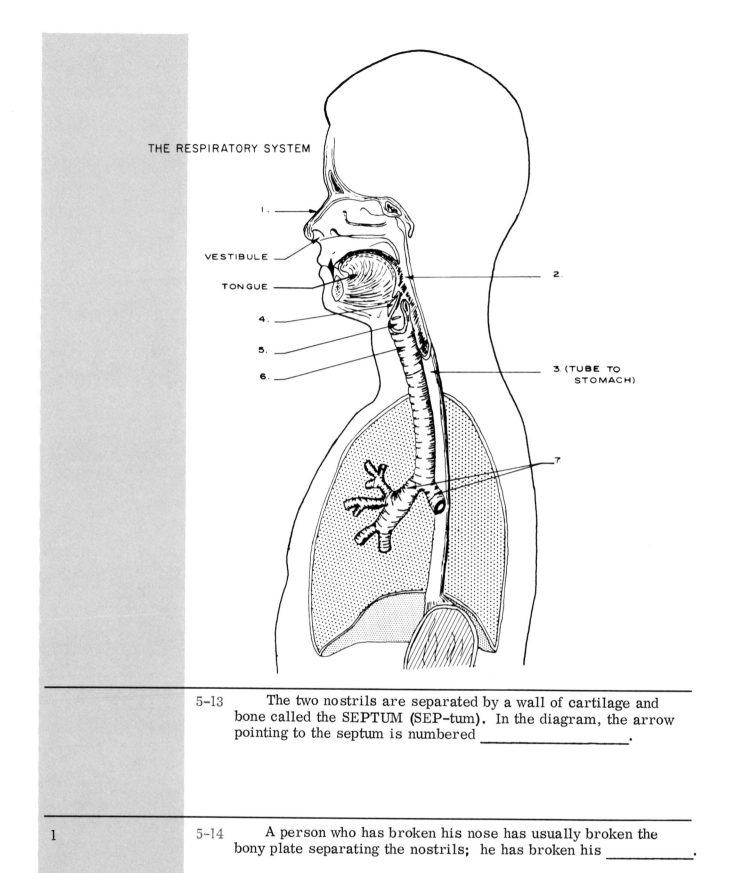

THE RESPIRATORY SYSTEM

1.

VESTIBULE

TONGUE

2.

4.

5.

6.

3. (TUBE TO STOMACH)

7.

5-13   The two nostrils are separated by a wall of cartilage and bone called the SEPTUM (SEP-tum). In the diagram, the arrow pointing to the septum is numbered _____.

1

5-14   A person who has broken his nose has usually broken the bony plate separating the nostrils; he has broken his _____.

140

septum	5-15	In the bone surrounding the nasal cavity are eight small air pockets called SINUSES (SY-nuh-sez). Since the sinuses have openings leading to the nasal cavity, their secretions drain into the _____ _____.
nasal cavity	5-16	Air pressure within the sinuses is equalized by the presence of passageways to the _____.
nasal cavity	5-17	Both air from the nostrils and food from the mouth pass into a single tube called the PHARYNX (FAR-inks). In the diagram, the arrow that points to the pharynx is numbered _____. (Refer to the diagram located between frames 12 and 13.)
2	5-18	The pharynx serves as a passageway for both _____ and _____.
air food (either order)	5-19	Food reaches the ESOPHAGUS (e-SOF-a·gus) after passing through the tube called the _____.
pharynx	5-20	The esophagus is the immediate extension of the pharynx; through it, food passes into the stomach. In the diagram, the arrow which points to the esophagus is numbered _____. (Refer to the diagram located between frames 12 and 13.)
3	5-21	While food passes directly through the pharynx into the esophagus, air takes a different route when it leaves the pharynx. In the diagram, the arrow numbered 4 points to the opening of the tube into which air passes. Air passes into a tube (in back/in front) _____ of the esophagus.

in front	5-22	Air passes through the LARYNX (LAR-inks) or "voice box", where vocal sounds originate. The front part of the larynx is the "Adam's apple." In the diagram, the arrow pointing to the larynx is numbered _____.
5	5-23	At the opening of the larynx is a disc of cartilage called the EPIGLOTTIS (ep-i-GLOT-is). In the diagram the arrow pointing to the epiglottis is numbered _____.
4	5-24	When food is swallowed, an automatic nerve muscle action closes the epiglottis and prevents food from entering the _____.
larynx	5-25	Occasionally, the automatic closure does not occur quickly enough, and food becomes lodged in the larynx, producing a choking sensation. Normally, however, food does not enter the larynx because the _____ closes to block its passage.
epiglottis	5-26	Label the numbered items in this drawing.

(1) septum (2) pharynx (3) esophagus (4) epiglottis (5) larynx	5-27	The tube through which both air and food pass is called the _____.
pharynx	5-28	The tube which both acts as an air passage and produces vocal sounds is called the _____.
larynx	5-29	The larynx sits on top of a tube called the TRACHEA (TRA-ke-a), which is commonly referred to as the "windpipe". In the diagram, the arrow pointing to the trachea is numbered _____.
6	5-30	The pharynx, larynx and trachea, like the nostrils, are lined with hair-like _____ ,which clean and filter incoming air.
cilia	5-31	In the chest, the trachea forks into a right and left BRONCHUS (BRON-kus), leading to the right and left lungs, respectively. In the diagram (Page 142), the arrow pointing to the bronchi (plural of bronchus) is numbered _____.
7	5-32	Air passes from the windpipe or _____ into right and left _____.

trachea bronchi	**5-33** The ridge where the bronchi begin is known as the CARINA (ka-RE-na). Irritation of the carina triggers the cough reflex. Thus the "cough reflex center" in the brain receives impulses from the _____ which is located at the juncture of the two _____.

TRACHEA——

BRONCHI——

CARINA

carina bronchi	**5-34** The trachea and bronchi are composed in part of cartilage rings. Within the lungs, the bronchi branch into the increasingly smaller and smaller bronchi. The smallest bronchi, barely visible without a microscope, still contain _____ rings.
cartilage	**5-35** The bronchi continue to branch into microscopic tubes which no longer contain cartilage but are composed of muscle tissue. These tiniest branches of the bronchi are called BRONCHIOLES (BRON-ke-olz); they differ from the bronchi in that they contain no _____ rings but are composed only of _____ tissue.

cartilage
muscle

5-36   Label the parts of this drawing.

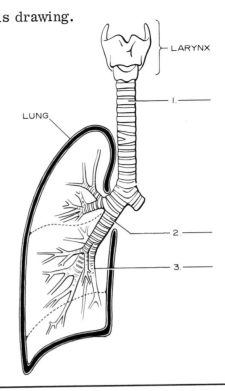

1. trachea
2. bronchus
3. bronchiole

5-37   Within the lungs, the bronchioles terminate in small air sacs called _____.

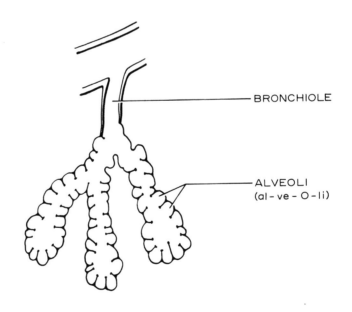

alveoli

5-38   Oxygen is carried through the blood stream to all parts of the body. Oxygen passes into the _____ stream from the air sacs called _____ in the lungs.

     (We shall discuss this process in more detail later in the program.)

blood alveoli	5-39	The alveoli are located in the primary organs of respiration, the two _____ .
lungs	5-40	The lungs are housed in the thoracic cavity. They are covered by a serous membrane (a membrane which secretes a watery fluid) called the PLEURA (PLOO-ra). The thoracic cavity is also lined by a _____ like that which covers the lungs.
pleura	5-41	Because it is lined by pleura, the _____ cavity is also called the PLEURAL CAVITY.
thoracic	5-42	The thoracic cavity is divided into a right and left cavity by a partition called the MEDIASTINUM (me-de-AS-ti-num). The mediastinum contains all the major organs of the thoracic cavity <u>except</u> the lungs. One lung is located on either side of the _____ .
mediastinum	5-43	The heart and great vessels, the trachea, part of the esophagus and many smaller structures are all part of the _____ .
mediastinum	5-44	The lungs are asymmetrical. The heart projects into the left thoracic cavity and creates a notch called the _____ _____ in the left lung.

RIGHT LUNG —

— LEFT LUNG

CARDIAC NOTCH

146

cardiac notch	5-45 Deep fissures in the lungs divide them into LOBES. In the right lung there are _____(how many?) lobes. In the left lung there are _____ lobes. 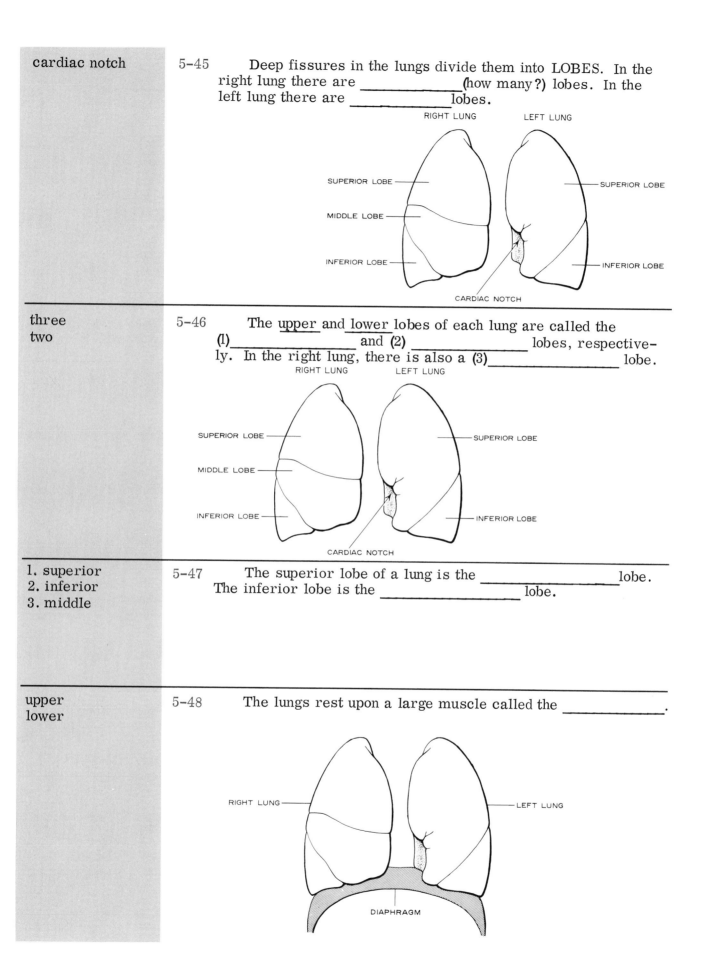
three two	5-46 The <u>upper</u> and <u>lower</u> lobes of each lung are called the (1)_____ and (2) _____ lobes, respectively. In the right lung, there is also a (3)_____ lobe.
1. superior 2. inferior 3. middle	5-47 The superior lobe of a lung is the _____ lobe. The inferior lobe is the _____ lobe.
upper lower	5-48 The lungs rest upon a large muscle called the _____ .

diaphragm	5-49	During <u>inspiration</u>, the diaphragm <u>contracts</u> and moves down, thus (increasing/decreasing) _____ the volume of the thoracic cavity.

LUNG
INSPIRATION
EXPIRATION
DIAPHRAGM
INSPIRATION
EXPIRATION

increasing	5-50	During expiration, the diaphragm relaxes and moves up, thus (increasing/decreasing) _____ the volume of the thoracic cavity.

LUNG
INSPIRATION
EXPIRATION
DIAPHRAGM
INSPIRATION
EXPIRATION

decreasing	5-51	The diaphragm moves down and increases the size of the thoracic cavity during _____. It relaxes and moves up to decrease the size of the thoracic cavity during _____.
inspiration expiration	5-52	You are probably aware that you can, to some extent, control respiration (by "holding your breath"). However, you do not normally have to think about breathing; it is an unconscious process. Thus, while some voluntary control of respiration is possible, respiration is normally (a voluntary/an involuntary) _____ process.
an involuntary	5-53	The volume of the thoracic cavity is <u>greater</u> during (inspiration/expiration) _____ than it is during (inspiration/expiration) _____.

inspiration expiration	5-54 Respiration results from pressure changes as well as from muscle action. <u>Pressure</u> is the <u>force</u> exerted upon a unit of a given surface. For example, <u>atmospheric pressure</u> is the _____ which the air or atmosphere exerts upon surfaces.
force	5-55 At sea level, atmospheric pressure is about <u>14.7 pounds</u> per square inch. Thus, on a window measuring 3 feet by 9 feet, atmospheric pressure exerts a force of more than <u>28 tons.</u> If, as is usually the case, the pressures on both sides of the window are equal, the window will not break. (No answer required)
	5-56 Air is a gas. Whenever possible, gases tend to move in a way that equalizes their pressures. If we have an <u>open</u> container, we can expect that the pressure inside the container will be (equal to/different from) _____ the pressure outside the container.
equal to	5-57 Hypoxia (hi-POX-ee-a) is the technical term for oxygen lack. The organ <u>most</u> sensitive to hypoxia is the _____, whose tissues can be permanently damaged by oxygen lack in three to five minutes.
brain	5-58 The organ whose sensitivity to anoxia is second only to that of the brain is the _____.
heart	5-59 Exhausting air from a container lowers its internal pressure and creates a partial vacuum. If we open the container again, air will rush in to fill the vacuum, and the internal pressure will again _____ the external pressure.
equal (be equal to)	5-60 A law of physics (Boyle's Law) states that in a <u>closed</u> space, at a constant temperature, the pressure of a gas varies <u>inversely</u> with its volume. Thus, when the volume of such a gas <u>decreases</u>, its pressure <u>increases.</u> And when the volume of such a gas increases, its pressure _____.

decreases	5-61 Suppose you blow some air into a balloon and close the neck of the balloon; then, with your hand, you force the air into a smaller and smaller area of the balloon. When you do this, you are (increasing/decreasing) _____ the volume of air in the balloon.

(1)   (2)   (3)

decreasing	5-62 If you continue to force the air into a smaller and smaller space, the increased internal pressure will cause the balloon to break, because, as you _____ the volume of air, you _____ the pressure.

(1)   (2)

decrease (d) increase (d)	5-63 The thorax is a closed space whose size, hence volume of air, is altered by movements of the large muscle called the _____.

diaphragm	5-64 When the diaphragm contracts and moves down during inspiration, the size of the thoracic cavity (decreases/increases) _____.

LUNG
INSPIRATION
EXPIRATION

DIAPHRAGM
INSPIRATION
EXPIRATION

150

increases	5-65 When the volume of the thorax increases during inspiration, the pressure within the thorax naturally:
	
	    (a) increases (go on to frame 66)
	    (b) decreases (go on to frame 67)

	5-66 You said that when the volume of the thorax increases, the pressure within the thorax increases. <u>Wrong.</u> The thorax is a closed space; so its pressure changes in <u>inverse</u> ratio with its volume. When the volume of the thorax <u>increases</u>, the pressure <u>decreases.</u> Conversely, when the volume <u>decreases</u>, the pressure _____. (Go on to frame 67).
increases	5-67 You said that when the volume of air in the thorax increases, the pressure decreases. <u>You're right.</u> Conversely, when the volume of air decreases, the pressure _____.
increases	5-68 The pressure within the thorax is called intrathoracic pressure, and it is usually <u>less</u> than atmospheric pressure. During expiration, intrathoracic pressure increases, but it never equals _____ _____. The only exceptions to this are during coughing or when one bears down, at which time the pressure is temporarily greater.
atmospheric pressure	5-69 Because it is usually less than atmospheric pressure, _____ pressure creates a slight vacuum within the thorax.
intrathoracic	5-70 Because intrathoracic pressure creates a slight vacuum, it exerts a suction on the elastic lungs, so that they never <u>fully</u> collapse but always remain somewhat expanded. When intrathoracic pressure <u>decreases</u>, the suction <u>increases</u>, and the lungs _____ still further.

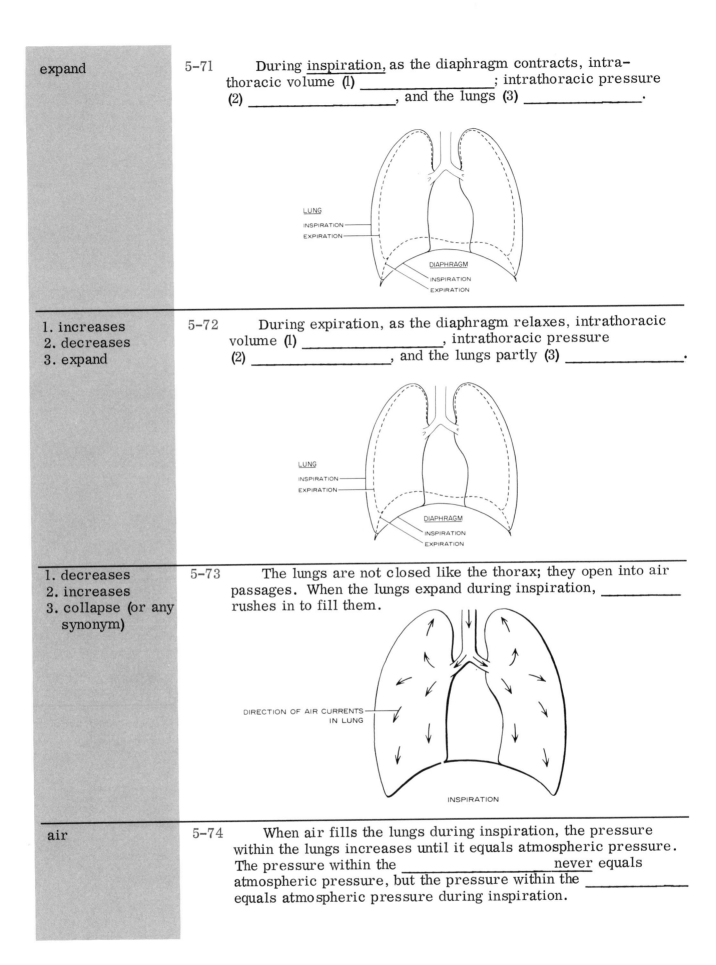

expand	5-71 During <u>inspiration</u>, as the diaphragm contracts, intrathoracic volume (1) _____; intrathoracic pressure (2) _____, and the lungs (3) _____.
1. increases 2. decreases 3. expand	5-72 During expiration, as the diaphragm relaxes, intrathoracic volume (1) _____, intrathoracic pressure (2) _____, and the lungs partly (3) _____.
1. decreases 2. increases 3. collapse (or any synonym)	5-73 The lungs are not closed like the thorax; they open into air passages. When the lungs expand during inspiration, _____ rushes in to fill them.
air	5-74 When air fills the lungs during inspiration, the pressure within the lungs increases until it equals atmospheric pressure. The pressure within the _____ never equals atmospheric pressure, but the pressure within the _____ equals atmospheric pressure during inspiration.

thorax lungs	5-75 During inspiration, intrathoracic volume (1) _____; intrathoracic pressure (2)_____, and the lungs (3) _____. At the same time, air rushes into the lungs, and the pressure in the lungs (4) _____. 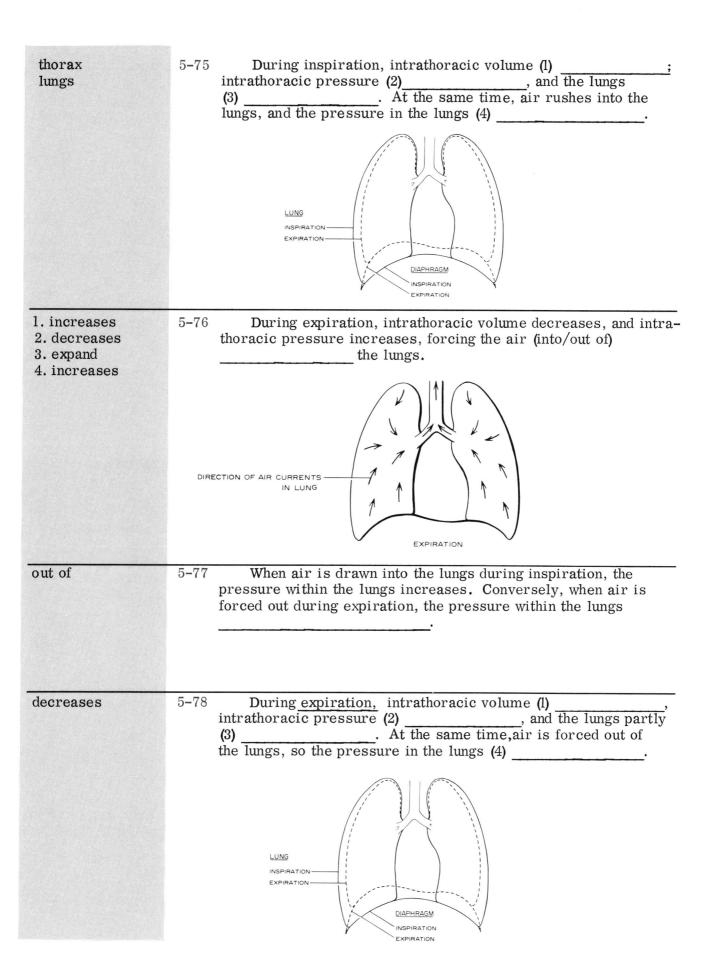
1. increases 2. decreases 3. expand 4. increases	5-76 During expiration, intrathoracic volume decreases, and intra- thoracic pressure increases, forcing the air (into/out of) _____ the lungs.
out of	5-77 When air is drawn into the lungs during inspiration, the pressure within the lungs increases. Conversely, when air is forced out during expiration, the pressure within the lungs _____.
decreases	5-78 During expiration, intrathoracic volume (1) _____, intrathoracic pressure (2) _____, and the lungs partly (3) _____. At the same time, air is forced out of the lungs, so the pressure in the lungs (4) _____.

1. decreases 2. increases 3. collapse 4. decreases	5-79 Describe the changes in volume and pressure within the thorax and lungs during _expiration._ (Use your own words).
Intrathoracic volume decreases; intrathoracic pressure increases; the lungs collapse, and the pressure within the lungs decreases.	5-80 Describe the changes in volume and pressure within the thorax and lungs during _inspiration._ (Use your own words).
Intrathoracic volume increases; intrathoracic pressure decreases; the lungs expand, and the pressure within the lungs increases.	5-81 Normal unstrained respiration is called QUIET RESPIRATION and consists of two phases: QUIET INSPIRATION and _____ _____.
quiet expiration	5-82. The amount of air inhaled and exhaled during quiet respiration averages about 500 cc. for the adult male. During one respiratory cycle only about 350 cc. of this reaches the alveoli, while the remaining air, called DEAD SPACE AIR, remains in the respiratory passages. Normally, the respiratory passages contain 150 cc. of _____ _____ _____.
dead space air	5-83 Dead space air is air which does not reach the _____, but remains in the _____ _____.
alveoli (lungs) respiratory passages	5-84 The amount of air inhaled and exhaled during each cycle of quiet respiration is called TIDAL VOLUME. For a normal adult male,_____ _____ averages 500 cc. of air.
tidal volume	5-85 Tidal volume is the amount of air exchanged per breath during _____ respiration.

quiet	5-86	With maximal effort, the quantity of air exchanged is greatly increased. The total amount of air that can be forcibly expelled from the lungs after a maximal inspiration is called the VITAL CAPACITY. Since the vital capacity of an adult male averages 4000 cc., vital capacity is (greater than/less than) _____ tidal volume, which is 500 cc.
greater than	5-87	The air which can be forcibly expelled from the lungs after a maximal inspiration is called the _____ _____.
vital capacity	5-88	During inspiration, some air remains in the respiratory passages as dead space air. The remainder of the inspired air passes into the air sacs called _____ in the lungs.
alveoli	5-89	Oxygen passes from the alveoli into the blood stream. The alveoli are enclosed by tissue, as are the blood vessels. Thus, to get from the alveoli into the blood stream, oxygen must pass through a wall of _____.
tissue(s)	5-90	When carbon dioxide passes out of the blood stream, it too must pass through _____.
tissue(s)	5-91	Oxygen and carbon dioxide pass through tissues by a process called diffusion. Thus we say that oxygen diffuses out of the alveoli into the blood stream and that carbon dioxide _____ out of the blood stream into the alveoli.

155

diffuses

5-92   Consider the mechanics of diffusion. Suppose we have a container like that in the illustration. The container is closed except for a gas hose at point A. If we pump gas into the container, the gas will <u>momentarily</u> be concentrated where it enters at point _____.

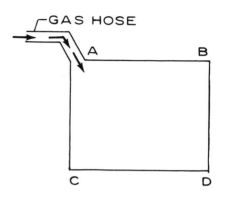

A

5-93   The "concentration" of a gas refers to the number of gas molecules present at a given point. When we say that a gas is more concentrated at point A than at point B, we mean that there are more _____ _____ present at point A than at point B.

gas molecules

5-94   Diffusion refers to the tendency of gas molecules to move from an area of high concentration to an area of low concentration. Thus if gas enters at point A and if gas molecules are briefly concentrated there, the molecules:

(a) will remain concentrated at A.

(b) will move throughout the container to B, C, D, and other points.

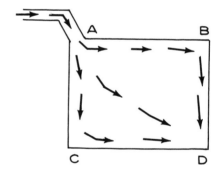

(b) will move throughout the container to B, C, D and other points.

5-95   The gas molecules in a given space will move until the concentration of molecules is equal at all points within the space. Thus, in a closed space, we (find/do not find) _____ a higher concentration of gas molecules at any one point than at another point.

do not find

5-96   A PERMEABLE MEMBRANE is a membrane through which gas molecules can pass. Because oxygen and carbon dioxide can pass through the tissues of the alveoli and blood vessels, these tissues are _____ membranes.

permeable

5-97   The principle of diffusion applies to gases separated by a permeable membrane. The illustration shows a container with two compartments separated by a permeable membrane. If gas molecules are concentrated on Side A, the molecules will tend to move through the _____ _____ to side

_____.

PERMEABLE
MEMBRANE

A   B

**(permeable)
membrane**

B

5-98   Suppose that in the illustrated container we have a gas which is a mixture of both oxygen (O$_2$) and carbon dioxide (CO$_2$). On side A there are more oxygen molecules than on side B. Although the gas is a mixture, oxygen molecules will still diffuse independently over to side _____.

PERMEABLE
MEMBRANE

A   B

O →

B

5-99   If the mixture of carbon dioxide and oxygen has a greater concentration of carbon dioxide molecules on side B than on side A, the _____ _____ molecules will diffuse independently over to side _____.

PERMEABLE
MEMBRANE

A   B

O →

← CO$_2$

**carbon
dioxide
A**

5-100   Thus we see that in a mixture of gases, the molecules of each gas diffuse independently. If one gas in a mixture has a higher concentration on side A of a membrane than on side B, the molecules of that gas will _____

_____.

(use your own words)

diffuse independently to side B, or move to the other side or any similar statement	**5-101** The human body constantly uses oxygen and gives off carbon dioxide. The body receives oxygen through the inspiration of air. Air is a mixture of gases which has a higher concentration of _____ than of carbon dioxide.
oxygen	**5-102** When blood enters the capillaries of the lungs, it has a low concentration of oxygen, while the air in the alveoli of the lungs contains a <u>high</u> concentration of oxygen. Therefore the oxygen diffuses out of the _____ into the _____.
alveoli (lungs) blood	**5-103** The blood entering the capillaries of the lungs has a high concentration of <u>carbon dioxide,</u> while the air in the alveoli has a low concentration of carbon dioxide. Therefore, the carbon dioxide diffuses out of the _____ into the _____
blood alveoli	**5-104** The transfer of oxygen into the blood and of carbon dioxide into the alveoli takes place by the process of _____.
diffusion	**5-105** Once the carbon dioxide from the blood reaches the alveoli, it traverses the same route followed by incoming air during inspiration. The carbon dioxide is forced up and out through the respiratory passages during (inspiration/expiration)_____.
expiration	**5-106** The oxygen which has passed into the bloodstream is carried to the cells throughout the body. The membranes surrounding these cells, like the membranes of the capillaries, are _____ membranes.
permeable	**5-107** When the oxygenated blood reaches the cells, there is a higher concentration of oxygen in the blood than in the cells, and a higher concentration of carbon dioxide in the cells than in the blood. Therefore, oxygen diffuses into the _____ and carbon dioxide diffuses into the _____.

cells blood	5-108	After giving up its oxygen to the cells, blood carries a high concentration of carbon dioxide and a low concentration of oxygen. The blood then returns to the alveoli of the lungs where, once more, _____ _____ diffuses out of the capillaries and _____ diffuses into the blood.
carbon dioxide oxygen	5-109	ARTERIES (AR-ter-eez) carry blood away from the heart and lungs. Therefore ARTERIAL (ar-TER-e-al) BLOOD is blood which is being carried: (a) away from the body cells toward the lungs (b) away from the lungs toward the body cells
(b) away from the lungs to- ward the body cells	5-110	Since arterial blood is travelling from the lungs to the body cells, it has a high concentration of (oxygen/carbon dioxide) _____.
oxygen	5-111	VEINS carry blood from the cells of the body back to the heart and lungs. VENOUS (VEE-nus) BLOOD, therefore, is blood which is being carried away from the (1)_____ back to the (2) _____ and (3)_____.
1. cells 2. heart 3. lungs	5-112	VENOUS blood has given up its oxygen to the cells and has, therefore, a (high/low) _____ concentration of oxygen. Venous blood also has a high concentration of carbon dioxide.
low	5-113	Blood contains a protein molecule called HEMOGLOBIN (he-mo-GLO-bin). It is this protein molecule which carries the oxygen in the blood. The term OXYHEMOGLOBIN (OX-e-he-mo-GLO-bin) refers to the molecule of _____ when it is combined with _____.
hemoglobin oxygen	5-114	Oxyhemoglobin is, of course, present in (arterial/venous) _____ blood.

arterial	5-115	When the hemoglobin molecule has given up its oxygen to the body tissue, it is called REDUCED HEMOGLOBIN. Reduced hemoglobin is present in (arterial/venous) _____ blood.
venous	5-116	Since tissue metabolism is perpetual, the body's production of carbon dioxide is also perpetual. The ACIDITY of the blood is in part dependent upon the total amount of carbon dioxide present in the blood. The condition known as ACIDOSIS (as-id-O-sis) results if the lungs become incapable of eliminating _____ _____.
carbon dioxide	5-117	The inability of the lungs to eliminate carbon dioxide results in _____.
acidosis	5-118	A condition known as ALKALOSIS (al-ka-LO-sis) results when the lungs eliminate too much _____.
carbon dioxide	5-119	When the lungs eliminate too much carbon dioxide, the resultant condition is _____. When they eliminate too little carbon dioxide, the resultant condition is _____.
alkalosis acidosis	5-120	Diseases which impair the ability of the lungs to transfer oxygen into the blood may produce HYPOXIA (hi-POX-e-a), in which too little oxygen enters the blood, or ANOXIA (a-NOX-e-a), in which no _____ enters the blood.

oxygen	**5-121** The regulation of the respiratory cycle is controlled by RESPIRATORY CENTERS in the brain. Respiration is thus controlled by the central nervous system. (no answer needed)
	5-122 The respiratory centers respond chiefly to the <u>carbon dioxide</u> content of the blood. An increase in the carbon dioxide content of the blood increases the rate of respiration while a decrease in the carbon dioxide content _____ the rate of respiration.
decreases	**5-123** Respiration depends chiefly upon the response of the respiratory centers to the (oxygen/carbon dioxide) _____ _____ content of the blood.
carbon dioxide	**5-124** A short review will help you remember what you have learned in this section. The outermost limits of the respiratory system are the nostrils. The vestibule or dilated portion of each nostril is lined with _____ ,which serve to _____ incoming air.
hairs filter	**5-125** Above the vestibule, the passages of the respiratory system are lined with hair-like _____ ,which also help to clean incoming air.

cilia

5-126   Label the numbered
parts of this drawing.

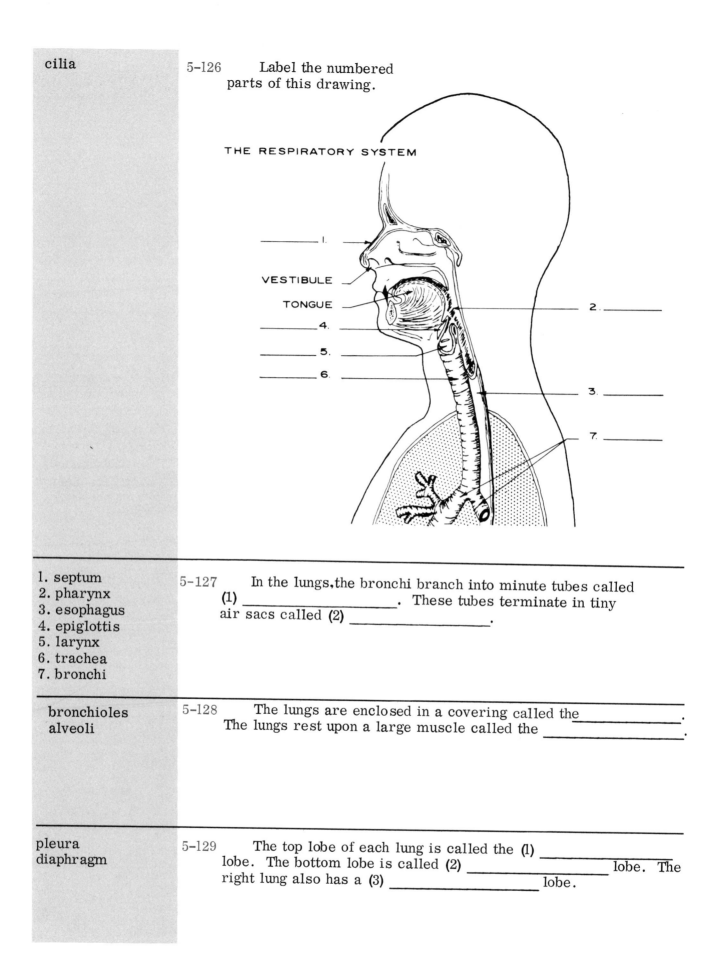

THE RESPIRATORY SYSTEM

VESTIBULE

TONGUE

1. septum
2. pharynx
3. esophagus
4. epiglottis
5. larynx
6. trachea
7. bronchi

5-127   In the lungs, the bronchi branch into minute tubes called
(1) _____. These tubes terminate in tiny
air sacs called (2) _____.

bronchioles
alveoli

5-128   The lungs are enclosed in a covering called the _____.
The lungs rest upon a large muscle called the _____.

pleura
diaphragm

5-129   The top lobe of each lung is called the (1) _____
lobe. The bottom lobe is called (2) _____ lobe. The
right lung also has a (3) _____ lobe.

162

1. superior 2. inferior 3. middle	5-130	During inspiration, the diaphragm (contracts/relaxes) _____. During expiration, the diaphragm (contracts/relaxes) _____.
contracts relaxes	5-131	During inspiration, volume of the thoracic cavity _____ and intrathoracic pressure _____.
increases decreases	5-132	During expiration, the volume of the thoracic cavity _____ and intrathoracic pressure _____.
decreases increases	5-133	During inspiration, the lungs (1) _____ as air is (drawn in/pushed out) (2) _____. During expiration, the lungs (3) _____ as air is (drawn in/pushed out) (4) _____.
1. expand 2. drawn in 3. collapse 4. pushed out	5-134	Intrathoracic pressure is always (equal to/less than) _____ atmospheric pressure.
less than	5-135	Oxygen passes into the blood stream from the _____ in the lungs.
alveoli	5-136	Oxygen and carbon dioxide pass through the tissue membranes of the alveoli and capillaries in a process called _____.

diffusion

5-137　　Diffusion occurs because gas molecules tend to pass from an area of _____ concentration to an area of _____ concentration.

high
low

5-138　　Oxygen in the blood is carried by a protein molecule called _____.

hemoglobin

5-139　　The rate of respiration depends chiefly on the amount of _____ in the blood.

carbon dioxide

5-140　　Now that you have completed Chapter 5, you will want to take a break before beginning Chapter 6, the Circulatory System.

CHAPTER

6

The Circulatory System

6-1.   The CIRCULATORY SYSTEM, the subject of this unit, is the transportation system of the body. It is composed of the CARDIOVASCULAR (CAR – de – o – VAS – kew – lar) SYSTEM, which circulates blood, and the LYMPHATIC (lim – FAT – ik) SYSTEM, which circulates LYMPH (LIMF). Both the cardiovascular and the lymphatic systems are members of the

_____ _____.

---

circulatory
system

6-2.   The larger part of this unit will be devoted to the cardiovascular system. The word "cardiovascular" can be broken into two parts: "cardio," meaning "heart," and "vascular," meaning "vessel." The cardiovascular system, as its name implies, includes the _____ and the blood _____.

---

heart

vessels

6-3.   The cardiovascular system transports _____ throughout the body.

---

blood

6-4.   The heart is the "pump" of the _____ system.

---

cardiovascular

6-5.   The system of blood vessels in the body is sometimes called the VASCULAR TREE. The analogy stems from the fact that from a single large vessel, the AORTA (a – OR – ta), grow branches which become progressively smaller until they terminate in tiny "twigs." The main trunk of the vascular tree is the large _____.

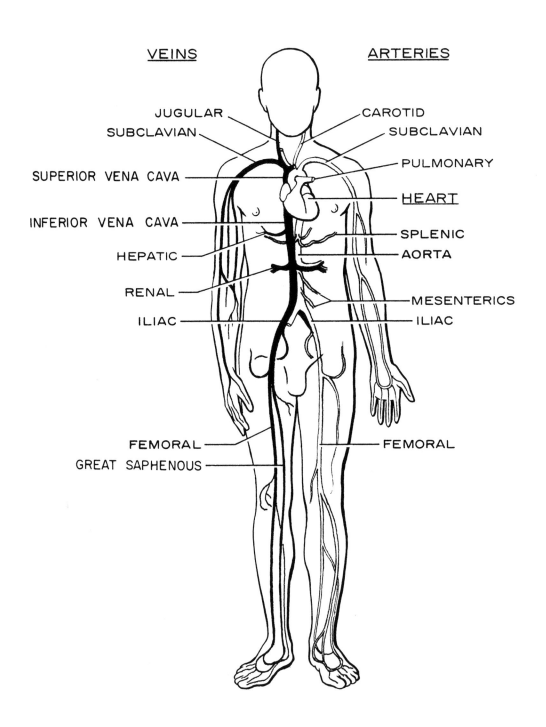

VEINS   ARTERIES

JUGULAR   CAROTID

SUBCLAVIAN   SUBCLAVIAN

PULMONARY

SUPERIOR VENA CAVA

HEART

INFERIOR VENA CAVA

SPLENIC

HEPATIC   AORTA

RENAL

MESENTERICS

ILIAC   ILIAC

FEMORAL   FEMORAL

GREAT SAPHENOUS

aorta	6-6. The smallest blood vessels in the body are called CAPILLARIES (CAP – il – a – reez). The capillaries are the "twigs" of the _____ _____.
vascular tree	6-7. The vascular tree has a double system of branches. The _____ carry blood <u>to</u> the heart. The _____ carry blood <u>away from</u> the heart.

VEINS ——— ARTERIES

HEART |
| veins

arteries | 6-8.   The trunk of the vascular tree, the aorta, carries blood away from the heart. Technically speaking, the aorta is a(n) _____. |
| artery | 6-9.   The connection between the arteries and the veins is accomplished by the "twigs" of the vascular tree, which are the _____. |
| capillaries | 6-10.   The walls of arteries are composed of three layers called TUNICS. The walls of veins are thinner than those of arteries but are also composed of three _____. (Use technical term) |

tunics	6-11. The innermost tunics of the veins and arteries are composed of a specialized epithelium called ENDOTHELIUM (en - do - THEE - le - um). The two outer tunics are missing from capillaries, which are composed only of _____ like that which lines the veins and arteries.
endothelium	6-12. Arteries are classified as SMALL, MEDIUM and LARGE. Large arteries are composed chiefly of elastic tissue while small and medium arteries are composed chiefly of muscular tissue. On the basis of the predominating tissue, small and medium arteries are called (muscular/elastic) _____ arteries, and large arteries are called (muscular/elastic)_____ arteries.
muscular elastic	6-13. The term "muscular arteries" refers to (1)_____ and (2)_____ arteries. The term "elastic arteries" refers to (3)_____ arteries.
(1) small (2) medium (3) large	6-14 The smallest arteries are called ARTERIOLES (ar-TER-e-olz). Arterioles, like larger arteries, have an inner tunic of endothelium and _____ (how many?) outer tunics.
two	6-15. The smallest arteries are called _____, and they, like other arteries, carry blood (to/away from) _____ the heart.
arterioles away from	6-16. Although the walls of veins are proportionately thinner than those of arteries, the veins, like the arteries, are composed of three _____.
tunics	6-17. Veins, like arteries, are classified as small, medium, and large veins and as VENULES (VEN-yewls), which are very (small/large) _____ veins.

small	6-18. Very small veins are called _____ , and they, like other veins, conduct blood (to/away from) _____ the heart.
venules to	6-19. The twigs of the vascular tree, the capillaries, do <u>not</u> have the outer two _____ found in the arteries and veins.
tunics	6-20. Lacking the outer tunics, the _____ are composed only of _____ like that which lines the rest of the cardiovascular system.
capillaries endothelium	6-21. As you know, the transfer of oxygen and carbon dioxide between the blood and tissues take place through the capillary walls. Nutrients and waste products are also transferred between the (1)_____ and the (2) _____ through the (3) _____ walls.
(1) blood (2) tissues (either order) (3) capillary	6-22. The total system of capillaries in a particular area is called the CAPILLARY BED. In order to meet the demands of all the body cells for food and oxygen, the capillary bed is extensive. It has been estimated that approximately 60,000 miles of capillaries form the _____ _____ of skeletal muscle alone.
capillary bed	6-23. A brief review. (1) _____ carry blood to the heart. (2) _____ carry blood away from the heart to all parts of the body. The tiniest blood vessels, through which food and oxygen pass into the tissues, are called (3) _____ .
(1) veins (2) arteries (3) capillaries	6-24. Both veins and arteries are composed of an inner lining of (1) _____ and two outer (2) _____ . Capillaries are composed only of (3) _____ .

(1) endothelium	6-25. The "pump" of the cardiovascular system is the
(2) tunics	_____ .
(3) endothelium	

---

heart

6-26.   The heart is composed chiefly of muscle. The muscle of the heart is called the MYOCARDIUM (my – o – KAR – de – um). A connective tissue sac encloses the heart muscle or

_____ .

---

myocardium

6-27.   Contractions of the heart muscle, called the _____ , pump blood throughout the body.

---

myocardium

6-28.   The heart has four chambers. The top chambers are the right atrium and the left _____ . The bottom chambers are the right and left _____ .

(Note: The drawings of the heart in this program are schematics designed to show the directions in which blood flows. They are not exact representations of the way the heart looks.)

RIGHT ATRIUM   LEFT ATRIUM

RIGHT VENTRICLE   LEFT VENTRICLE

---

atrium

ventricles

6-29.   The plural of "atrium" is "atria." In the heart, the two atria are located (above/below)_____ the two ventricles. The upper chambers of the heart are also referred to as "auricles."

above	6-30. The major pumping force of the cardiovascular system comes from the lower two chambers of the heart, called the _____.
ventricles	6-31. The <u>arrows</u> in this and the following drawings represent the direction in which <u>blood</u> circulates through the heart. Blood from the body enters the <u>right atrium</u> through two veins called the _____ _____ _____ ____ and the _____ _____ _____.

SUPERIOR VENA CAVA

INFERIOR VENA CAVA

RIGHT ATRIUM

RIGHT VENTICLE

superior vena cava inferior vena cava	6-32. Blood from the body passes through the superior vena cava into the _____ _____ of the heart.
right atrium	6-33. "Superior" means "above," and "inferior" means "below." These terms, applied to the venae cavae (plural of vena cava), describe the areas drained by the two veins. Blood from the upper part of the body drains into the _____ _____ _____, while blood from the lower part of the body drains into the _____ _____ _____.
superior vena cava inferior vena cava	6-34. The superior and inferior venae cavae drain all of the body <u>except</u> the lungs. Thus the blood which passes through the venae cavae into the right atrium has given up its oxygen to the tissues and is (rich/poor) _____ in oxygen.

poor	6-35 Between the right atrium and right ventricle is a valve called the _____. When this valve opens, blood passes from the right atrium into the _____.

RIGHT ATRIUM —
TRICUSPID VALVE —
RIGHT VENTRICLE —

---

tricuspid valve right ventricle	6-36. Blood from the body enters the (1) _____ _____ of the heart. The blood then passes through the (2) _____ valve into the (3) _____ _____ of the heart.

---

(1) right atrium (2) tricuspid (3) right ventricle	6-37 A powerful contraction of the right ventricle forces blood through the _____ valve into the _____, _____ which carries blood to the lungs.

PULMONARY ARTERY —
PULMONIC VALVE —

---

pulmonic pulmonary artery	6-38. Blood from the right ventricle enters the lungs by passing through the _____ valve into the _____ _____, which leads to the lungs.

---

pulmonic pulmonary artery	6-39. Blood which has passed through the pulmonary artery into the capillaries of the lungs gives off _____ _____ and receives _____ in the process of respiration.

carbon dioxide oxygen	6-40　Oxygen-rich blood from the lungs enters the <u>left</u> atrium of the heart through two veins, one for each lung, called the _____.

PULMONARY VEINS
LEFT ATRIUM
LEFT VENTRICLE

pulmonary veins	6-41.　Blood from the lungs passes through the pulmonary veins into the _____ _____ of the heart. The blood entering this chamber of the heart is (rich/poor) _____ in oxygen.

left atrium rich	6-42.　Between the left atrium and the left ventricle is a valve called the _____. When this valve opens, blood passes through it into the _____ of the heart.

LEFT ATRIUM
MITRAL VALVE
LEFT VENTRICLE

mitral valve left ventricle	6-43.　Blood from the lungs enters the _____ _____ of the heart by passing through the two _____.

left atrium pulmonary veins	6-44.　Blood from the left atrium then passes through the _____ valve into the _____ of the heart.

mitral left ventricle	6-45. A powerful contraction of the left ventricle then pushes the blood from the heart into the main artery of the body, the _____. LEFT ATRIUM AORTA LEFT VENTRICLE
aorta	6-46. Oxygen-rich blood from the heart reaches the entire body through branches of the main artery called the _____.
aorta	6-47. Let's review the location of the valves in the heart. The tricuspid valve is between the (1)_____ and the (2) _____. The mitral valve is between the (3)_____ _____ and the (4) _____. The pulmonic valve is between the (5)_____ artery and the (6)_____ _____. TRICUSPID VALVE — MITRAL VALVE PULMONIC VALVE —
(1) right atrium (2) right ventricle (3) left atrium (4) left ventricle (5) pulmonary (6) right ventricle	6-48. Now, a brief review of the circulation of blood through the heart. Oxygen-poor blood from all of the body except the lungs enters the (1) _____ _____ of the heart through the (2)_____ _____ _____ and the (3) _____ _____ _____.

174

(1) right atrium (2) superior vena cava (3) inferior vena cava	6-49	Blood from the right atrium passes through the _____ valve into the _____ _____ of the heart.
tricuspid right ventricle	6-50	A contraction of the right ventricle pushes the blood through the (1) _____ valve into the (2) _____ _____, which carries the blood to the (3) _____.
(1) pulmonic (2) pulmonary artery (3) lungs	6-51	After oxygenation in the lungs, blood re-enters the heart by passing through the two _____ _____. The blood from the lungs enters the _____ _____ of the heart.
pulmonary veins left atrium	6-52	Oxygen-rich blood from the left atrium passes through the (1) _____ valve into the (2) _____ of the heart. A contraction of the heart then forces the blood out into the (3) _____.
(1) mitral (2) left ventricle (3) aorta	6-53	Oxygen-rich blood is carried throughout the body by arteries, all of which are branches of a main artery called the _____.
aorta	6-54	The brain is very richly supplied with blood. The richness of the blood supply to the brain reflects the importance of a constant supply of oxygen to the brain. Since brain tissue may be permanently damaged by oxygen deprivation in as short a time as three to five minutes, it is clear that _____ is of immense importance to the _____.
oxygen brain	6-55	Because it can be permanently damaged by oxygen lack in as little as three to five minutes, the _____ is the organ which is most sensitive to oxygen deprivation.

brain

6-56   The tissues of the heart are nourished by blood from two major arteries, the RIGHT and LEFT CORONARY ARTERIES. In its need for a constant supply of oxygen, the _____ is second only to the brain.

heart

6-57   Consider what happens when we change the pressure inside a container. To begin with, we have an open container, so the pressure inside it is _____ _____ the pressure outside it.

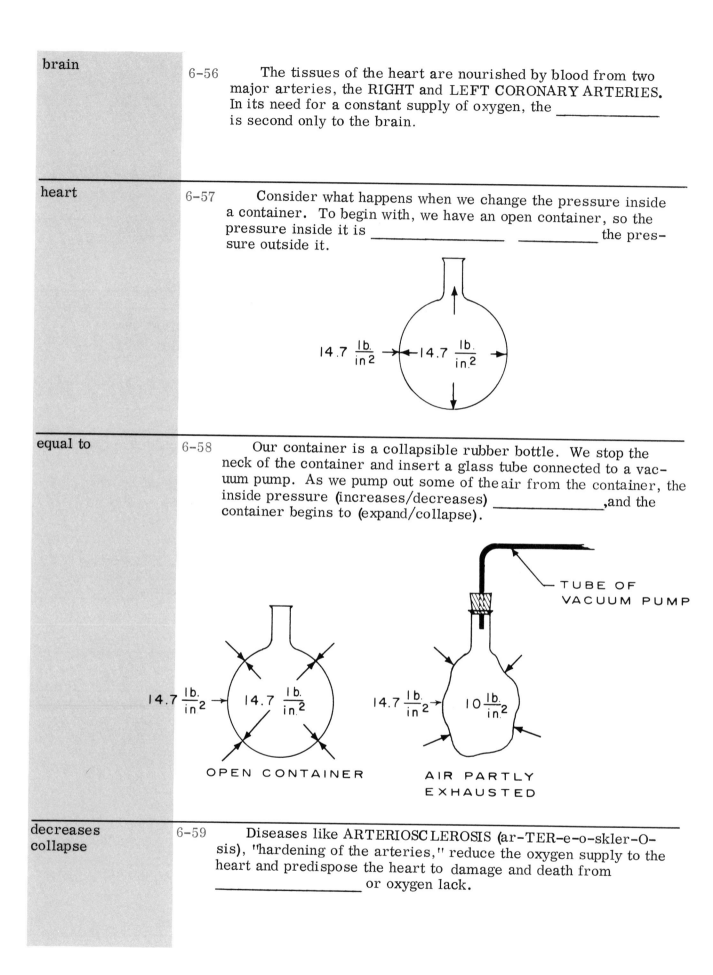

equal to

6-58   Our container is a collapsible rubber bottle. We stop the neck of the container and insert a glass tube connected to a vacuum pump. As we pump out some of the air from the container, the inside pressure (increases/decreases) _____ ,and the container begins to (expand/collapse). _____

TUBE OF
VACUUM PUMP

OPEN CONTAINER

AIR PARTLY
EXHAUSTED

decreases
collapse

6-59   Diseases like ARTERIOSCLEROSIS (ar-TER-e-o-skler-O-sis), "hardening of the arteries," reduce the oxygen supply to the heart and predispose the heart to damage and death from _____ or oxygen lack.

hypoxia	6-60	Like the arteries that supply the brain, the coronary arteries are interconnected so that an obstruction in one artery may not result in death from anoxia. In the event of such an obstruction, the entire myocardium may (but does not always) still receive adequate blood through the unimpaired _____ artery.

coronary	6-61	The myocardium is drained by a <u>venous</u> system (a system of veins) that more or less parallels its <u>arterial</u> system. Similarly, the other parts of the body receive blood through _____ systems and are drained by nearly parallel _____ systems.

arterial venous	6-62	You have already learned about the circulation of blood through the heart. The following frames will introduce you to the mechanisms which control this circulation. (No answer needed).

	6-63	The circulation of blood is controlled by the alternate contraction and relaxation of the heart muscle, called the _____.

myocardium	6-64	The alternate contraction and relaxation of the myocardium constitutes the CARDIAC CYCLE. Blood circulation is maintained by the _____ _____.

cardiac cycle	6-65	The cardiac cycle consists of two phases. The period during which the myocardium contracts is called _____. The period during which the myocardium relaxes is called _____.

<u>CARDIAC CYCLE</u>

SYSTOLE
(CONTRACTION)

DIASTOLE
(RELAXATION)

systole diastole	6-66	SYSTOLE (SIS-to-le), during which the myocardium (1) _____, and DIASTOLE (di-AS-to-le), during which the myocardium (2) _____, are the two phases of the (3) _____ _____.

VENTRICULAR SYSTOLE
ATRIAL DIASTOLE

ATRIAL SYSTOLE
VENTRICULAR DIASTOLE

(1) contracts (2) relaxes (3) cardiac cycle	6-67	The chambers of the heart do not go into systole simultaneously. ATRIAL SYSTOLE occurs when the _____ contract. VENTRICULAR SYSTOLE occurs when the _____ _____.
(1) atria (2) ventricles contract	6-68	When the ventricles contract, the atria relax; when the atria contract, the ventricles relax. During atrial systole, the ventricles (contract/relax) _____ or go into diastole.
relax	6-69	During diastole, the chambers of the heart are resting and preparing for the pumping effort of systole. While they relax, they are filling with blood. Blood enters the chambers of the heart when they (contract/relax) _____ and is pumped out when they (contract/relax) _____.
relax contract	6-70	The chambers of the heart fill with blood during _____ and pump out the blood during _____.

178

diastole systole	6-71 Systole is initiated by the spread of an electrical impulse from a specialized node of tissue called the CARDIAC PACEMAKER. The cardiac pacemaker is located in the _____ _____ of the heart. CARDIAC PACEMAKER
right atrium	6-72 The node of tissue which initiates atrial systole is called, in technical terminology, the SINO - ATRIAL (SI-no-A-tre-al) NODE, or simply the S A NODE. The _____ _____ node is also referred to as the _____ pacemaker.
S A cardiac	6-73 Systole begins when an electrical impulse is transmitted by the (1) _____ _____ node, also called the (2)_____ _____, which is located in the (3) _____ _____ of the heart.
(1) S A (2) cardiac pacemaker (3) right atrium	6-74 When the electrical impulse initiating systole leaves the S A node or cardiac pacemaker, the <u>first</u> chamber of the heart to go into systole is the chamber in which the cardiac pacemaker is located, namely the _____ _____.

179

right atrium	6-75	Within .02 of a second, the impulse spreads from the right atrium to the left atrium, which then goes into systole.

Although the two atria do not contract simultaneously, their contractions occur so close together that they can be discussed as if they contracted at the same time. Atrial systole, involving both atria, takes only about .08 of a second. The term "atrial systole" refers to the contraction of both the _____ and the _____ atria.

right left (either order)	6-76	During the period of diastole which precedes atrial systole, the atria relax and fill with blood. Thus, at the beginning of atrial systole, both atria are filled with _____.

blood	6-77	At the beginning of atrial systole, the right atrium is filled with oxygen-poor blood from the _____, while the left atrium is filled with oxygen-rich blood from the _____.

body lungs	6-78	As the two atria contract in atrial systole, the oxygen-poor blood in the right atrium is pumped into the _____ _____, and the oxygen-rich blood in the left atrium is pumped into the _____.

right ventricle left ventricle	6-79	When the atria go into systole and pump their blood into the two ventricles, the ventricles are relaxing; that is, they are in _____.

diastole

6-80 When the electrical impulse which originated at the S A node or cardiac pacemaker has traversed the two atria, it spreads to another node, called the ATRIO-VENTRICULAR (A-tre-o-ven-TRIK-u-lar) NODE, or simply the A V NODE. The A V node, like the S A node, is located in the _____ _____.

SA NODE —

AV NODE —

— BUNDLE OF HIS

— VENTRICULAR SEPTUM

---

right atrium

6-81 The atrio-ventricular, or A V node, is located in the right _____,but it transmits impulses to the two lower chambers of the heart, the _____.

---

atrium
ventricles

6-82 Systole begins when an electrical impulse is transmitted from the cardiac pacemaker or _____ _____ node. When this impulse has traversed the atria, it reaches the _____ _____ node,which transmits the impulse to the ventricles.

S A
A V

6-83 At the A V node, the impulse is delayed while the ventricles receive whatever blood continues to enter from the atria. The impulse is then transmitted to the specialized conducting system of the ventricles,and the ventricles contract or go into _____.

181

systole	6-84	Separating the right from the left ventricle is a wall of muscle called the _____ _____.

SA NODE

AV NODE — BUNDLE OF

VENTRICULA SEPTUM

ventricular septum	6-85	The impulse from the A V node is carried through the ventricular septum to the walls of the ventricles by two branches of the _____ _____.

SA NODE

AV NODE — BUNDLE OF

VENTRICUL SEPTUM

bundle (of) His	6-86	When the impulse transmitted by the bundle of His reaches the ventricles, the ventricles contract in systole. Since the ventricular systole is preceded by ventricular diastole, the ventricles are filled with _____ when they go into systole.

blood	6-87	As the ventricles go into systole, the atria go into _____ and begin to fill with blood.

diastole	6-88	When the ventricles go into systole, the right ventricle contains (oxygen-rich/oxygen-poor) _____ blood from the body, and the left ventricle contains (oxygen-rich/oxygen-poor) _____ blood from the lungs.

oxygen-poor oxygen-rich	6-89	During ventricular systole, the oxygen-poor blood in the right ventricle is pumped up through the pulmonary artery and goes to the _____ to pick up oxygen.
lungs	6-90	During ventricular systole, the oxygen-rich blood in the left ventricle is pumped out through the _____ to be carried to the cells of the body.
aorta	6-91	After ventricular systole, the ventricles again relax, or go into _____. At the same time, the two atria begin to go into _____.
diastole systole	6-92	Atrial systole begins when an electrical impulse is transmitted by the _____ _____ node or _____ _____.
S A cardiac pace- maker	6-93	The impulse originated by the S A node is transmitted next to the _____ _____ node where it is temporarily delayed.
A V	6-94	During atrial systole, blood is pumped from the _____ into the _____.
atria ventricles	6-95	During atrial systole, when blood from the atria is being pumped into the ventricles, the ventricles are in _____.

diastole	6-96	At the end of atrial systole, the atria go into _____ , and the impulse which has been delayed at the A V node is transmitted to the _____ _____ _____ , which is the conducting system of the ventricles.
diastole bundle of His	6-97	The impulse transmitted to the ventricles by branches of the bundle of His causes the _____ to go into _____ .
ventricles systole	6-98	During ventricular systole, blood from the right ventricle is pumped into the _____ , and blood from the left ventricle is pumped into the _____ .
pulmonary artery (lungs) aorta (body)	6-99	The ability of the heart to supply the body's need for blood depends partly upon its ability to fill with an adequate blood supply during diastole. Without an adequate period of rest in diastole, the heart cannot receive an adequate blood supply and cannot, therefore, supply adequate _____ to the body.
blood	6-100	Hypoxia and other diseases which produce abnormally high rates of heart action naturally shorten the period of diastole, and, after a critical level, tend to (increase/decrease) _____ the heart's intake and output of blood.
decrease	6-101	Some diseases, such as rheumatic fever, may damage the impulse-conducting system of the heart. Damage to the A V node may cause the atria and ventricles to act independently and uncooperatively. Here, too, the ventricles may begin to contract before they are fully filled with blood, and the body (will/will not) _____ receive an adequate supply of blood.
will not	6-102	Efficient functioning of the cardiovascular system is obviously of vital importance to the health of the body. The phrase "the cardiovascular system" implies only one system, but scientists recognize _two_ separate systems for the circulation of blood. (No answer required)

184

	6-103 As you know, arteries carry blood away from the heart. Most of the arteries in the body are filled with blood which has just been oxygenated in the lungs and has, therefore, a (high/low) _____ oxygen content.
high	6-104 The pulmonary artery carries oxygen-poor blood from the right ventricle to the lungs. Therefore, the pulmonary artery differs from the other arteries in that it carries oxygen-_____ blood instead of oxygen-_____ blood.
poor rich	6-105 Most of the veins in the body carry oxygen-poor blood back to the heart. The pulmonary vein, however, transports blood from the lungs into the left atrium. Unlike other veins, the pulmonary vein carries oxygen-_____ blood.
rich	6-106 The term ARTERIAL BLOOD refers to oxygen-rich blood. The term VENOUS BLOOD refers to oxygen-poor blood. Most arteries in the body transport (arterial/venous) _____ blood, but the pulmonary artery leading to the lungs transports (arterial/venous) _____ blood.
arterial venous	6-107 Most veins in the body carry _____ blood, but the pulmonary vein leading from the lungs to the left atrium carries _____ blood.
venous arterial	6-108 On the basis of the difference in the oxygen content of the blood in veins and arteries, the cardiovascular system is subdivided into two systems. Thus, the veins and arteries of all of the body except the lungs are considered to be members of one system, while the veins and arteries of the _____ are considered to be members of another system.
lungs	6-109 SYSTEMIC (sis-TEM-IK) CIRCULATION is the circulation of blood through all of the _____ except the lungs. PULMONIC CIRCULATION is the circulation of blood through the _____.

body lungs	6-110	The head and feet, for instance, receive blood through _____ circulation. The lungs receive blood through _____ circulation.
systemic pulmonic	6-111	The terms "blood pressure" and "pulse" are frequently used in connection with the cardiovascular system. The following frames will be devoted to a consideration of these terms. (No answer required.)
	6-112	When the <u>left</u> ventricle contracts, blood pulses through the arteries. Certain large arteries are located close to the surface of the skin, and the pulsation of blood through these _____ can be felt through the skin.
arteries	6-113	"Pulse" refers to the pulsation of _____ through large arteries. The pulse can be felt when the _____ _____ contracts and forces blood through the arteries.
blood left ventricle	6-114	Although it is commonly felt through the radial artery in the wrist, the _____ can be felt in any part of the body where an _____ is close to the skin.
pulse artery	6-115	If the <u>left</u> ventricle is not filled with blood when it contracts, little blood will be forced through the arteries, and the pulse will be weak/strong_____.
weak	6-116	Pulse depends partly upon the presence of an adequate supply of _____ in the _____ _____ when it contracts.

186

blood left ventricle	6-117	If there is an obstruction between the left ventricle and the artery in which the pulse is sought, the pulse will be weak or absent. Therefore, the pulse also depends upon the absence of _____ between the left ventricle and the _____ in which the pulse is to be felt.
obstruction artery	6-118	If the left ventricle does not contract with sufficient force to pump all of its blood into the arteries, the pulse will be weak. Pulse also depends upon the amount of _____ with which the _____ _____ contracts.
force left ventricle	6-119	Pulse depends upon: 1. The presence of an adequate (1) _____ supply in the (2)_____ _____ 2. The absence of (3) _____ between the (4) _____ _____ and the (5) _____ in which the pulse is sought. 3. The (6) _____ with which the (7) _____ _____ contracts.
1. blood 2. left ventricle 3. obstruction 4. left ventricle 5. artery 6. force 7. left ventricle	6-120	The cardiovascular system contains a series of blood vessels through which blood is pumped from the heart. "Blood pressure" is the pressure of the _____ within these blood _____.
blood vessels	6-121	The total diameter of all the capillaries in the capillary bed fed by an artery is much larger than the diameter of the artery itself. For this reason, the blood pressure in the arteries is higher than the pressure in the _____.
capillaries (capillary bed)	6-122	Blood pressure is higher in the arteries than it is in the capillaries. Blood pressure (is/is not) _____ the same in all the blood vessels of the body.

is not	6-123 Usually, when physicians measure blood pressure, they measure the pressure in the arteries. Therefore, the term "blood pressure" when used alone refers to the pressure in the _____.
arteries	6-124 When the left ventricle contracts during systole, ARTERIAL PRESSURE (blood pressure in the arteries) rises. When the left ventricle relaxes in diastole, the arterial pressure diminishes. Arterial pressure is <u>higher</u> during ventricular (systole/diastole) _____ than during ventricular (systole/diastole) _____.
systole diastole	6-125 SYSTOLIC PRESSURE is blood pressure during ventricular systole. DIASTOLIC PRESSURE is blood pressure during ventricular diastole. In any individual, _____ pressure is always <u>lower</u> than _____ pressure.
diastolic systolic	6-126 The textbook figure for the normal blood pressure of a resting young adult is 120/80. Naturally, the higher figure, 120, is a measure of _____ pressure, and the lower figure, 80, is a measure of _____ pressure.
systolic diastolic	6-127 If the left ventricle does not contract forcefully enough to pump all of its blood into the arteries, the blood pressure will be low. Blood pressure, like the pulse, depends partly upon the _____ with which the _____ _____ contracts.
force left ventricle	6-128 The walls of the arteries are somewhat elastic. This elasticity permits them to expand and recoil as blood pulses through them. If the arteries "harden," they lose their elasticity, and the blood pressure rises above normal. Blood pressure, then, depends partly upon the <u>elasticity</u> of the walls of the _____.
arteries	6-129 Two forces which influence blood pressure are: 1. The (1) _____ with which the (2) _____ _____ contracts, and 2. The (3) _____ of the walls of the (4) _____.

1. force 2. left ventricle 3. elasticity 4. arteries	6-130	Small arteries, arterioles, are capable of expanding and constricting so as to change the size of their lumen (central cavity). When the arterioles constrict and resist the blood flow, blood pressure in the arteries rises. When the arterioles relax, the blood flows through more freely, and blood pressure in the arteries (increases/decreases)_____.
decreases	6-131	The resistance provided by the arterioles influences the pressure of blood in the arteries. Thus, another factor influencing blood pressure is the _____ of the _____ to the flow of blood from the arteries.
resistance arterioles	6-132	The "viscosity" of blood refers to its fluidity (relative to water). Thick or viscous fluids flow more slowly than thin, aqueous fluids and require more pressure to move them through tubes. Blood pressure, therefore, is partly dependent upon the _____ of the _____.
viscosity (thickness) blood	6-133	Two factors which influence blood pressure are: 1. The resistance of the (1) _____ to the flow of blood from the arteries, and 2. The thickness or (2) _____ of the (3) _____.
(1) arterioles (2) viscosity (3) blood	6-134	No significant pressure can be achieved within a system of tubes if the system is not filled with fluid. If a great deal of blood is lost from the vascular system (through hemorrhage, for instance), the blood pressure will drop. Blood pressure depends also upon the volume of _____ in the vascular system.
blood	6-135	Blood pressure is influenced by: 1. The (1) _____ with which the (2) _____ _____ contracts. 2. The (3) _____ of the walls of the (4) _____. 3. The resistance of the (5) _____ to the flow of blood from the arteries.
(1) force (2) left ventricle (3) elasticity (4) arteries (5) arterioles	6-136	Blood pressure also depends upon: 1. The _____ of blood. 2. The _____ of blood.

volume viscosity (either order)	**6-137** Infection, drugs, and nervous influences may decrease the tone of the arterioles so that they offer too <u>little</u> resistance to blood from the arteries. In this case, blood will flow too freely through the arteries, and blood pressure will be (high/low) _____.
low	**6-138** During an accident in which a large amount of blood is lost, the blood volume decreases, and the blood pressure (increases/decreases) _____.
decreases	**6-139** Constriction of the arterioles may very gravely affect the blood pressure, producing <u>ARTERIAL HYPERTENSION</u>. In arterial hypertension, the constricted arterioles prevent blood from flowing through the arteries. Thus, the blood volume in the arteries increases, and blood pressure (rises/drops) _____.
rises	**6-140** The following frames will be devoted to a discussion of blood, its composition and its functions. An average-size adult male weighing 150 lbs. has approximately 5 quarts of blood in his body. (No answer needed)
	6-141 Blood is composed of a group of cells (plus fluid), each fulfilling specific functions, and all functioning together. Blood, therefore, like skin and bone, is a _____.
tissue	**6-142** Blood functions primarily in carrying food and the vital gas _____ to all the cells of the body and in removing from the cells the waste products of metabolism, such as the waste gas _____ _____.
oxygen carbon dioxide	**6-143** When the body is exposed to cold, the arterioles near the skin help to maintain body temperature by constricting, thus reducing the surface area of blood from which heat may be lost. When the body is exposed to heat, these same arterioles expand to produce a larger surface flow of blood from which _____ may be lost.

heat	6-144	Thus, one function of the blood, assisted by the vessels which carry it, is to maintain body _____.
temperature	6-145	Blood has other indispensable functions; it carries water, electrolytes, and hormones to the tissues. Blood serves as the transportation system of the body, transporting not only oxygen and carbon dixoide, but also (1) _____, (2) _____ and (3) _____.
(1) water (2) electrolytes (3) hormones	6-146	More than half of the blood is a watery fluid called PLASMA (PLAZ-ma), and the remainder consists of formed elements which can be viewed under a microscope. The formed elements are classified as (1) _____ _____ _____, (2) _____ _____ _____ and (3) _____ _____.

BLOOD PLATELET

RED BLOOD CELL

WHITE BLOOD CELLS

(1) red blood
    cells
(2) white blood
    cells
(3) blood plate-
    lets
	6-147	The capillaries through which red blood cells travel are often smaller than the cross-sectional diameters of the red blood cells themselves. An internal elastic framework maintains the shape of the cells but permits them to bend and twist as they pass through the small _____.

capillaries	6-148	The red blood cells contain a protein molecule called HEMOGLOBIN (HE-mo-glo-bin), which has an affinity for oxygen. Because they contain _____, the red blood cells are able to carry _____ to the body cells.
hemoglobin oxygen	6-149	The red color of the blood is produced by the protein molecule, _____, carried in the _____ _____ _____.

191

hemoglobin red blood cells	6-150	Approximately 2-1/2 million new red blood cells enter the blood every second, and an equal number are lost every second. The continual production of red blood cells is the function of the red _____ contained in the bones.
marrow	6-151	Red blood cells are atypical cells because they have no nucleus. White blood cells, on the other hand, contain organoids, inclusions and a more or less central nucleus. Thus, white blood cells are (typical/atypical) _____ cells.
typical	6-152	_____ blood cells are typical nucleated cells. _____ blood cells are atypical cells which have no nuclei.
white red	6-153	White blood cells appear to be largely inactive in the blood stream. Most of their activities are carried on outside the blood stream in the tissues of the skin and other organs. When body tissues are invaded by bacteria, the white blood cells enter the diseased _____ and ingest the _____.
tissue bacteria	6-154	Some white blood cells produce ANTIBODIES, which can combine with invading foreign proteins. The antibodies help to fight infection. Thus, the white blood cells counteract disease in two ways: they ingest _____, and they produce _____.
bacteria antibodies	6-155	Red blood cells are called ERYTHROCYTES (er-ITH-ro-sites), and white blood cells are called LEUKOCYTES (LU-ko-sites). _____ ingest bacteria and produce antibodies. _____ carry oxygen in the blood stream.
leukocytes erythrocytes	6-156	Use technical terms to answer this: _____ carry hemoglobin and are atypical cells without nuclei. _____ are typical nucleated cells which are inactive in the blood stream but active in the body tissues.

erythrocytes leukocytes	6-157	Blood platelets are also present in the blood. Their function is to plug leaks in injured blood vessels by forming blood clots. When a blood vessel is injured, certain components of the blood cooperate to form a mesh at the site of the injury. Blood _____ become entrapped in the mesh and form a blood _____ ,which plugs the leak.
platelets clot	6-158	Blood platelets are called THROMBOCYTES (THROM-bo-sites), and the clot which they form is called a THROMBUS (THROM-bus). When blood vessels are injured, the platelets or _____ cooperate with other elements in the blood to form a _____ or clot.
thrombocytes thrombus	6-159	Use technical terms to complete this item. (1) _____ carry oxygen in the blood stream. (2) _____ ingest bacteria and form anti-bodies. (3) _____ form blood clots.
(1) erythrocytes (2) leukocytes (3) thrombocytes	6-160	The blood of some individuals differs from that of others by virtue of its containing certain FACTORS. One such factor is the Rh FACTOR,which is present in the blood of some individuals and absent in others. Blood which contains the Rh factor is typed as Rh POSITIVE, and blood which does not contain the Rh factor is typed as _____ _____.
Rh negative	6-161	Blood is also classified according to the presence of the A and B factors. If blood contains the A factor, it is classified as TYPE A blood. If it contains the B factor, it is classified as TYPE B blood. Blood which contains both the _____ and the _____ factors is classified as TYPE A B blood.
A B	6-162	Blood which contains neither the A nor the B factors is classified as Type O blood. Blood is "typed" by accounting for the A and B factors and the Rh factor. If blood contains the A factor but not the Rh factor, it is classified as Type A, Rh negative. If blood contains neither the A nor B factors but does contain the Rh factor, it is classified as Type _____, _____ _____.
O, Rh positive	6-163	If a person with one type of blood receives a transfusion from a donor with another type of blood, the result is an INCOMPATIBLE TRANSFUSION,which can cause the death of the recipient. If someone with Type A blood receives a transfusion from a donor with Type B blood, the result is an _____ _____.

193

incompatible transfusion	6-164 An incompatible transfusion, which may cause the _____ of the recipient, results from the transfusion of blood between a donor who has one type of blood and a recipient who has (the same/a different) _____ type of blood.
death a different	6-165 The remainder of the program will be devoted to a discussion of the LYMPHATIC SYSTEM which, together with the cardiovascular system, forms the _____ system of the body.
circulatory	6-166 In the unit on the respiratory system, you learned that gas molecules will move through a permeable membrane, moving from an area of high concentration to an area of low concentration. The process by which the gas molecules move in this way is called _____.
diffusion	6-167 By a similar process, called OSMOSIS (os-MO-sis), water molecules can move through a membrane. The membrane through which water molecules pass in the process of osmosis is called a SEMIPERMEABLE MEMBRANE because some <u>large</u> particles dissolved in the water will not pass through it, but _____ molecules and some <u>small</u> dissolved particles will pass through it.
water	6-168 Let us illustrate osmosis. To begin with, we have a tank divided into two sections (A and B) by a _____. The tank contains plain water. The water level on both sides of the membrane (is/is not) _____ the same.
semipermeable membrane is	6-169 Now we add some sugar to the water on side B of the membrane, but not to the water on side A. The water level on side B of the membrane (rises/drops) _____ because (sugar/water) _____ molecules pass through the membrane from side A to side B.

rises water	6-170 When a semipermeable membrane separates two solutions containing different concentrations of dissolved particles, water molecules tend to pass through the membrane in an attempt to equalize the concentrations of the particles on both sides of the membrane. The process involved is called osmosis and refers to the passage of _____ molecules through a _____ membrane.
water semipermeable	6-171 As the capillaries traverse the tissues of the body, water and electrolytes osmose through the semipermeable membrane of the capillaries into the spaces between the cells of the _____.
tissues	6-172 The resulting fluid is called TISSUE FLUID. Tissue fluid is formed when, by the process of osmosis, _____ passes through the walls of the _____ into spaces between the cells of the tissue.
water capillaries	6-173 Tissue fluid helps to nourish tissues. It loses its oxygen and nutrients into the tissues and acquires waste products from the tissues. The tissue fluid itself then becomes a _____ product.
waste	6-174 Tissue fluid which has given up its nutrients and oxygen to the tissues and has acquired the waste products of metabolism then passes (by osmosis) into the LYMPHATIC (lim-FAT-ik) VESSELS. The tissue fluid combines with other elements in the lymphatic vessels to form LYMPH (limf). The lymphatic vessels carry the _____ back to its origin in the _____ stream.
lymph blood	6-175 Since lymph carries away the waste products of tissue metabolism, lymph is <u>not</u> carried <u>to</u> the tissues; it is only carried _____ from the tissues.
away	6-176 The drainage of lymph through the _____ vessels is a one-way flow, <u>not</u> a circulation.

lymphatic	6-177	The walls of the large arteries and veins are tissues. They are nourished partly by tissue fluid which, as a waste product, must be drained away. Hence, the walls of large (1) _____ and (2) _____ are drained by a system of (3) _____ vessels.
(1) veins (2) arteries (either order) (3) lymphatic	6-178	The lymphatic system originates as small blind lymphatic capillaries in the skin. These capillaries course centrally as they converge and form progressively larger vessels which ultimately empty into the great (arteries/veins) _____, which carry blood back to the heart.

veins	6-179	All of the organs of the body, with the exception of the central nervous system, the eye, and inner ear possess a lymphatic system. In which of these is a lymphatic system absent?
		
		  a. the walls of the lungs
		  b. the eye
		  c. the walls of arteries

b. the eye	6-180	Interspersed along the course of the lymphatic system are a series of glands known as LYMPH NODES. At one point, called the HILUS (HI-lus), the wall of the lymph node is indented. Through the hilus, blood vessels and nerves enter and leave the _____ _____.
lymph node (s)	6-181	Lymph nodes remove dust, carbon, bacteria and degenerating material from the lymph that passes through them. In somewhat the same way that hairs and cilia are the filters of the respiratory system, so lymph nodes are the _____ of the lymphatic system.
filters	6-182	LYMPHOCYTES (LIM-fo-sitz) are special cells produced by the lymphatic system. Lymphocytes are poured into the lymph as it is filtered by the _____ _____.
lymph nodes	6-183	Lymph nodes are widely distributed throughout the body. Some of them are close to the skin and can be felt through the skin. In response to disease, particularly infectious diseases, the _____ _____ may become swollen and tender.

lymph nodes	6-184 Located in the abdomen is the largest of all lymphoid organs, the SPLEEN, which filters blood, not lymph. Although it resembles a large lymph node, the _____ differs from the lymph nodes in that it filters _____.
spleen blood	6-185 At this point you may take a rest. This is the last item of the chapter on the circulatory system which, as you have learned, is composed of two systems, the _____ system and the _____ system.
cardiovascular lymphatic (either order)	

CHAPTER

7

Part 1 -
The Nervous System

7-1. You will recall that the structural and functional unit of nerve tissue--the neuron--was described in some detail in Chapter Three of this course. A review of nerve tissue is important to the understanding of how impulses are received and transmitted by your body's communication center, the _nervous_ system.

---

nervous

7-2. A neuron is composed of a cell body and fibers which can vary in length from less than an inch to several yards. There are two types of process--fiber--extending from the neuron's cell body: the axon and the dendrites. Label the two extensions of the cell body's cytoplasm:

2. _dendrite_

CELL BODY

1. _axon_

---

1. axon

2. dendrite

7-3. The axon is a fiber along which nerve impulses move <u>away</u> <u>from</u> the cell body. Dendrites, which are also fibers, conduct impulses _toward_ the cell body.

---

toward

7-4. Axons may be encased in:
1. A sheath of fatty <u>non-cellular</u> material-<u>MYELIN</u>
2. A sheath of <u>cellular</u> material (the <u>NEUROLEMMA</u> sometimes called <u>SCHWANN</u> sheath)
3. Or a <u>combination</u> of both sheaths.

When an axon is encased by both sheaths, the neurolemma is outermost and is separated from the axon by the non-cellular _myelin_ sheath.

198

7-5.

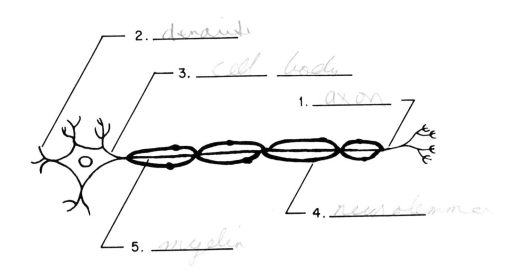

2. _denandi_

3. _cell body_

1. _axon_

4. _neurolemma_

5. _myelin_

Label the blank spaces.

1. axon
2. dendrite
3. cell body
4. neurolemma
   (Schwann
   sheath)
5. myelin

7-6.

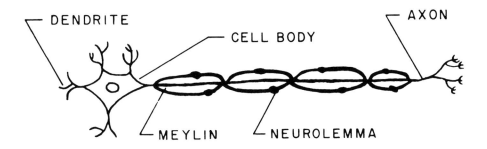

DENDRITE

CELL BODY

AXON

MEYLIN

NEUROLEMMA

The cellular axon insulation is the _neurolemma_, or Schwann sheath, and the _myelin_ is a fatty non-cellular insulating material.

neurolemma myelin	7-7. 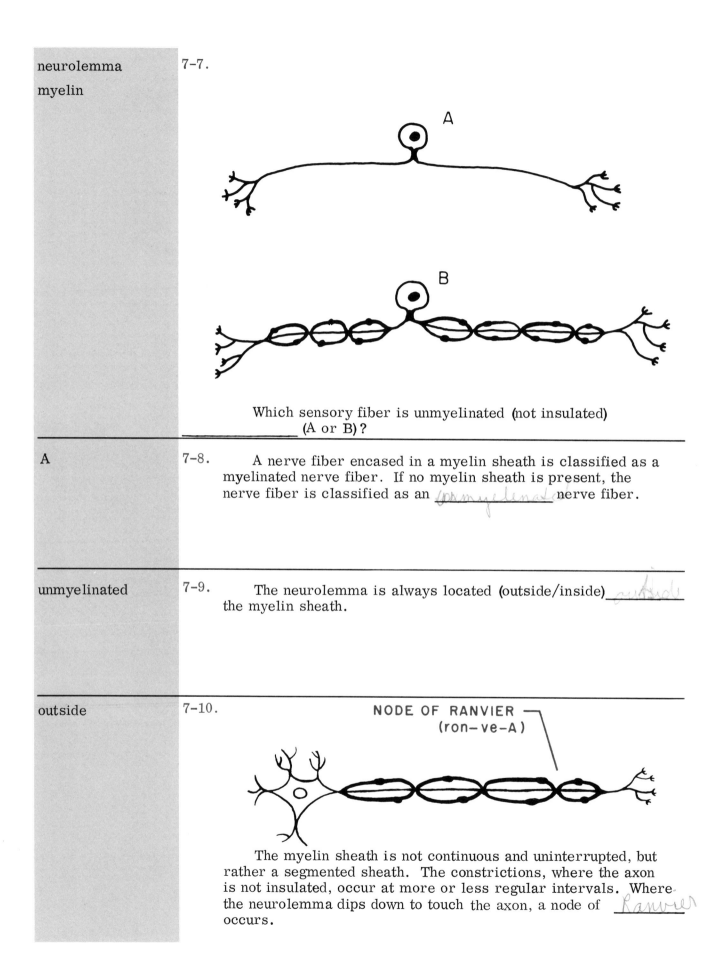 Which sensory fiber is unmyelinated (not insulated) _____ (A or B)?
A	7-8. A nerve fiber encased in a myelin sheath is classified as a myelinated nerve fiber. If no myelin sheath is present, the nerve fiber is classified as an *unmyelinated* nerve fiber.
unmyelinated	7-9. The neurolemma is always located (outside/inside) *outside* the myelin sheath.
outside	7-10. NODE OF RANVIER (ron-ve-A) The myelin sheath is not continuous and uninterrupted, but rather a segmented sheath. The constrictions, where the axon is not insulated, occur at more or less regular intervals. Where the neurolemma dips down to touch the axon, a node of *Ranvier* occurs.

Ranvier	7-11.	The portion of the nerve fiber that is far away--distant-- from its cell body uses myelin for nourishment. However, a nerve fiber that is distant from its cell body could be nourished by the surrounding fluids at the constrictions or _nodes of Ranvier_ .

nodes of Ranvier	7-12.	A nerve is many nerve fibers combined in a single bundle. This bundle of fibers, a ____nerve____, may contain both sensory and motor fibers.

nerve	7-13.	Fibers that receive and then conduct impulses from a sensory organ, the skin for example, to the central nervous system are called AFFERENT (AF-er-ent) fibers or sensory neurons. Fibers that conduct impulses from the central nervous system to the receptor organ, a muscle for example, and incite motor activity are called EFFERENT (EF-er-ent) fibers or_motor_ neurons.

motor	7-14.	In Chapter three of this book, reference was made to fibers that transmitted sensory impulses as _afferent_ fibers and to fibers that transmitted motor impulses as _efferent_ fibers.

afferent efferent	7-15.	Afferent fibers that transmit impulses from the periphery of the body to the brain or spinal cord have their cell bodies located outside the central nervous system. These fibers transmit (sensory/motor) _sensory_ impulses.

sensory	7-16.	Efferent fibers that transmit impulses to the periphery of the body from the brain or spinal cord have their cell bodies located within the central nervous system. These fibers transmit (sensory/motor) _motor_ impulses.

motor	7-17.	A NERVE NUCLEUS is a group of neuron or nerve cell bodies within the central nervous system. The cell bodies of _efferent_ fibers are part of a nerve _nucleus_ .

motor (efferent) nucleus	7-18. A <u>GANGLION</u> (GANG–gli–un) is a group of nerve cell bodies outside the central nervous system. Cell bodies of _afferent_ fibers can combine to form a _ganglion_ .

1. sensory
    (afferent)
 	
2. ganglion
	7-19. A group of nerve cell bodies within the central nervous system is:
	
	    (a) a ganglion (move to frame 20)
	
	    (b) a nucleus (move to frame 21)

(b)	7-20. No, a ganglion is a group of neuron cell bodies outside the central nervous system. The cell bodies of afferent fibers combine to form a ganglion. (Move to frame 21)
	7-21. Yes, a nucleus is a group of cell bodies within the central nervous system. The cell bodies of (motor/sensory)_____ neurons combine to form a nerve nucleus.
motor	7-22. An <u>involuntary</u> movement which is excited in response to a stimulus applied to a _sensory_ neuron and transmitted from the central nervous system via a motor neuron is a reflex. A reflex would depend upon the reception of a sensory impulse and transmission of the impulse <u>from</u> the central nervous system to the muscle via one or more _motor_ neurons.

sensory motor	7-23. When the afferent impulse from the skin or tendon travels to the central nervous system, it crosses a _____. The impulse then travels to an extensor muscle, inducing involuntary contraction. This is an example of a _____. 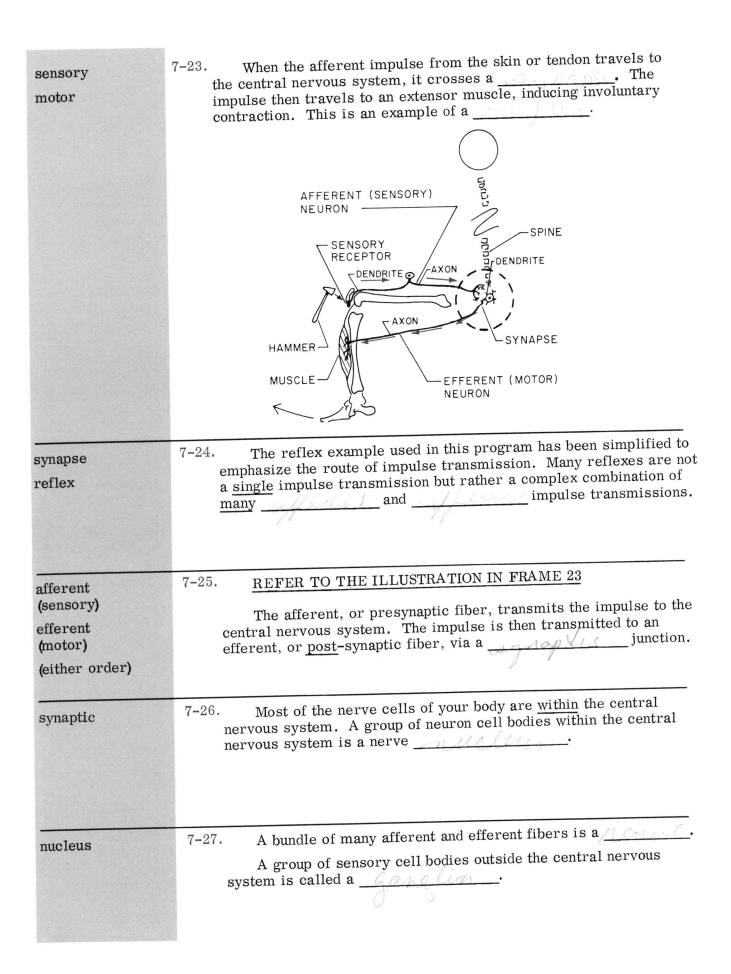

AFFERENT (SENSORY) NEURON

SENSORY RECEPTOR

DENDRITE

AXON

SPINE

DENDRITE

HAMMER

AXON

MUSCLE

SYNAPSE

EFFERENT (MOTOR) NEURON

synapse reflex	7-24. The reflex example used in this program has been simplified to emphasize the route of impulse transmission. Many reflexes are not a <u>single</u> impulse transmission but rather a complex combination of many _____ and _____ impulse transmissions.
afferent (sensory) efferent (motor) (either order)	7-25. <u>REFER TO THE ILLUSTRATION IN FRAME 23</u> The afferent, or presynaptic fiber, transmits the impulse to the central nervous system. The impulse is then transmitted to an efferent, or <u>post</u>-synaptic fiber, via a _____ junction.
synaptic	7-26. Most of the nerve cells of your body are <u>within</u> the central nervous system. A group of neuron cell bodies within the central nervous system is a nerve _____.
nucleus	7-27. A bundle of many afferent and efferent fibers is a _____. A group of sensory cell bodies outside the central nervous system is called a _____.

nerve ganglion	7-28.	The part of a fiber that is far away from its cell body may receive nourishment from the non-cellular _myelin_ or from the fluids surrounding the fiber which are in more direct contact with the axon at the constrictions or nodes of _Ranvier_.
myelin Ranvier	7-29.	A presynaptic fiber of a simple reflex arc is an _afferent_ fiber. A post-synaptic fiber of a simple reflex arc is an _efferent_ fiber.
afferent efferent	7-30.	For convenience and clarity, nerve tissue is divided grossly into three separate systems: The Central Nervous System (CNS) The Autonomic Nervous System (ANS) The Peripheral Nervous System Although this is a classic gross description, all nerve tissue is functionally integrated. (No response required—move to frame 31)
	7-31.	**CENTRAL NERVOUS SYSTEM** The central nervous system consists of the BRAIN and SPINAL CORD, which are housed in the skull --cranial vault -- and vertebral column respectively. Afferent impulses, which are transmitted first to the (1) _spinal cord_ may then either go to the (2) _brain_ or directly to an effector organ (muscle) via (3) _efferent fiber_.
1. spinal cord 2. brain 3. efferent fibers	7-32.	The central nervous system, CNS, consists of both nerve cells and nerve fibers--afferent and efferent--which conduct impulses between the principle parts of the CNS, the _brain_ and _spinal cord_, before the impulses arrive at the end of their travels.
brain spinal cord	7-33.	The brain and spinal cord are separated from the bony elements of the cranial vault and vertebral column by three fibrovascular membranes, the <u>MENINGES</u> (men-IN-jes). The central nervous system receives protection and support from the fibrovascular membranes, the _meninges_, the cranial _vault_, and the vertebral column.

meninges vault	7-34. A fresh brain removed from the cranial vault is soft and tends to lose its shape because of the forces exerted by its own weight. Within the cranial vault, the brain receives support from the _____ and a watery fluid between the fibrovascular membranes.
meninges	7-35. The outer <u>MENINX</u> (singular of meninges), the <u>DURA MATER</u> (DU-ra - MA-ter), lies immediately next to the inner surface of the skull and vertebral column. The brain and spinal cord are separated from the bony inner surfaces of the skull and vertebral column by the outer meninx, the _dura_ _mater_.
dura mater	7-36. The space within your skull -- cranial vault -- is not occupied by one large unsegmented "glob" of brain. The brain is divided into <u>communicating</u> segments; each segment is protected and supported by the folds of the outer meninx, the _dura_ mater.
dura	7-37. A second protective membrane, the <u>ARACHNOID</u> (a-RAK-noid), is separated from the dura mater by a watery fluid. The middle of the three protective meninges is the _arachnoid_.
arachnoid	7-38.

CRANIAL BONE

c.s.f.

DURA MATER

BRAIN

(fill in)

c.s.f.

PIA MATER

    Your brain is supported and protected by three membranes. The outer meninx, the (1)_dura_ mater; the mid-meninx, the (2) _arachnoid_ ; and the inner meninx the (3)_pia mater_.

1. dura 2. arachnoid 3. pia mater	7-39. A watery fluid containing water and electrolytes--the CEREBROSPINAL (SER-e-bro-SPI-nal) fluid--is located between the outer fibrovascular membrane, the (1) _____dura_____ mater, and the arachnoid. Cerebrospinal (2) _____fluid_____ is also located between the arachnoid and the inner fibrovascular membrane, the (3) _____pia_____ mater.
1. dura 2. fluid 3. pia	7-40. The three membranes--the dura, arachnoid, and pia mater-- and the (1) _____CS_____ fluid provide (2) _____protection_____ and (3) _____support_____ for the brain and spinal cord.
1. cerebrospinal 2. protection 3. support	7-41. The pia mater is a highly vascular membrane which closely conforms to the irregular brain surface. Blood vessels extend into the substance of the brain and spinal cord from the (dura/pia) _____pia_____ mater.
pia	7-42. At the tip of the spinal cord, the pia mater extends as a thread-like structure <u>through</u> the arachnoid and dura mater and attaches to the coccyx of the spine. The area between the pia mater and arachnoid is occupied by _____CSF_____ , which can circulate from the brain to the tip of the _____spinal cord_____ .
cerebrospinal fluid spinal cord	7-43 CRANIAL VAULT BRAIN 1. _____dura_____ MATER 2. _____arachnoid_____ 4. _____CS_____ FLUID 3. _____pia_____ MATER SPINAL CORD Fill in the blank spaces

206

1. dura
2. arachnoid
3. pia
4. cerebrospinal
	7-44. For convenience the following terms will be abbreviated in this program, as they are in medicine:
	
	    Central Nervous System .... CNS
	    Autonomic Nervous System .... ANS
	    Cerebrospinal fluid .... c.s.f.
	    (No response required—move to frame 45)

	7-45. The space between the arachnoid membrane and the pia mater is referred to as the subarachnoid space. Cerebrospinal fluid, (1) _C . S . F .,_ is located in the space between the arachnoid and the pia mater, the (2) _subarachnoid_ space, and in the space between the dura mater and the arachnoid, the (3) _sub_ dural space.

1. c.s.f. 2. sub- arachnoid 3. sub	**7-46.** Located within the brain and spinal cord are cavities which also contain c.s.f. The c.s.f. within these cavities of the brain and spinal cord (balance/unbalance) _balance_ the effect of the c.s.f. on the outer surface of the brain.

balance	**7-47.** The cavities <u>within</u> the brain which contain c.s.f. are called VENTRICLES (VEN-tri-kls), and the cavity within the spinal cord which contains c.s.f. is called the CENTRAL CANAL. The c.s.f. is produced in the cavities within the brain called _ventricles._

ventricles	**7-48.** C.s.f. is located in the subarachnoid and subdural spaces and also within the brain and the spinal cord in the _ventricle_ and the _central canal._

ventricles central canal	7-49. If the brain and spinal cord were dissected as illustrated in the drawing below, you could locate the c.s.f., brain cavities and the spinal canal.

LATERAL VENTRICAL

ONE SIDE OR HEMISPHERE OF THE BRAIN

3rd VENTRICAL

1. 4th _____
(fill in)

2. _____ CANAL
(fill in)

SPINAL CORD

Fill in the blanks.

1. ventricle 2. central	7-50. The c.s.f. that is formed in the _____ flows to the subarachnoid space.

ventricles	7-51. Some of the c.s.f. flows upward over the surface of the brain, and some of the c.s.f. flows down in the _____ space and the _____ _____ of the spinal cord.

subarachnoid central canal	7-52. The fluid pressure created by the c.s.f. on the outer surface of the brain is balanced by the c.s.f. fluid pressure in the _____ of the brain and _____ _____ of the spinal cord.

ventricles central canal	**7-53.** If the ventricles of the brain continued to produce c.s.f. and if the body had no functional means of reabsorbing the excess, increased pressure would result. There are within the cranial vault three major venous sinuses which function in part by absorbing c.s.f., thereby preventing any increase in _____.
pressure	**7-54.** Note that the c.s.f. is passing from the subarachnoid space through _____ _____ into a _____ _____.

SCALP

ARACHNOID VILLI

PARIETAL BONE

DURA MATER

VENOUS SINUS (sinus of venous blood)

PIA MATER

BRAIN   BRAIN

c.s.f.

ARACHNOID

FISSURE

arachnoid villi venous sinus	**7-55.** After the c.s.f. has been circulated over the surfaces of the brain and the spinal cord, the fluid is then absorbed through the _____ _____ into the _____ _____.
arachnoid villi venous sinus	**7-56.** The c.s.f. within and on the surface of the brain and spinal cord provides a balanced amount of _____ and _____ for these organs.
support protection	**7-57.** You may want to take a break before you continue. The next series of frames will deal with the brain.

7-58.

THE BRAIN

The human brain is divided into two hemispheres called CEREBRAL (SER-e-bral) hemispheres. The combined weight of the two _cerebral_ hemispheres is about 1350 grams or approximately 3 pounds for the average adult human.

---

cerebral

7-59.

The surface of each cerebral hemisphere is irregular or convoluted. A deep furrow is called a _fissure_ and a shallow indentation, a _sulcus_.

- LONGITUDINAL FISSURE SEPARATING EACH HEMISPHERE
- SULCUS (SUL-cus)
- GYRUS (JI-rus)
- SPINAL CORD

---

fissure

sulcus

7-60.

Refer to the illustration in frame 59.

A prominent rounded elevation on the surface of a cerebral hemisphere is called a _gyrus_.

---

gyrus

7-61.

Note: The plural of sulcus is SULCI, and the plural of gyrus is GYRI.

(No response required—move to frame 62)

7-62.

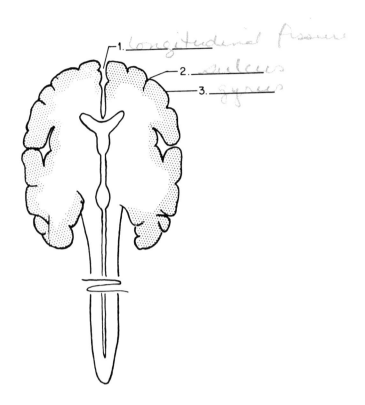

1. _longitudinal fissure_
2. _sulcus_
3. _gyrus_

Fill in the three blanks.

---

1. longitudinal
   fissure

2. sulcus

3. gyrus

7-63.   Each cerebral hemisphere has four main subdivisions or LOBES. In the illustration below locate each lobe.

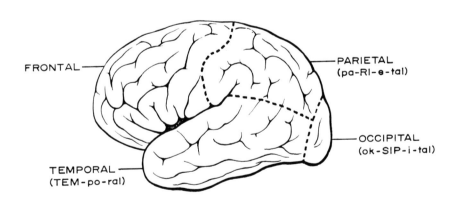

No answer required. Move to frame 64.

7-64.   The front and temple sections of your skull are familiar to you. Label the two subdivisions or (1) _____*lobes*_____ of the cerebral hemisphere illustrated below.

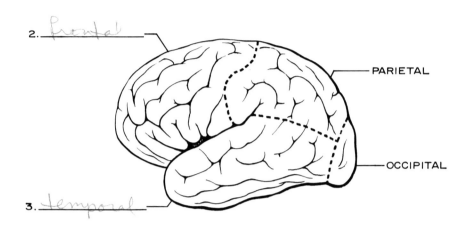

2. *frontal*

PARIETAL

OCCIPITAL

3. *temporal*

7-65.   Label the four lobes of the cerebral hemisphere illustrated below.

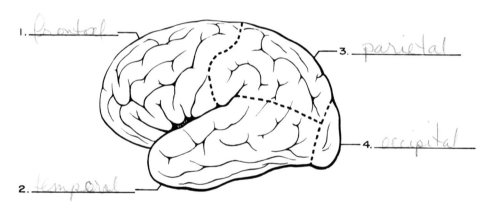

1. *frontal*

3. *parietal*

4. *occipital*

2. *temporal*

7-66.   Each cerebral lobe has a different functional activity. For example, the frontal lobe is a coordinating center for moral activities, the temporal lobe--hearing, the parietal (pa-RI-e-tal) lobe--touch, and the occipital (oc-SIP-i-tal) lobe--sight.

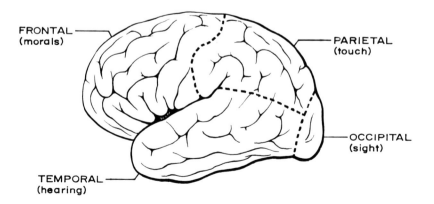

FRONTAL
(morals)

PARIETAL
(touch)

OCCIPITAL
(sight)

TEMPORAL
(hearing)

212

7-67.

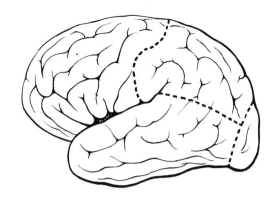

Match the cerebral lobe areas in the left hand column with the functional activities in the right hand column.

a. Frontal　　　　1. Touch

b. Occipital　　　2. Sight

c. Temporal　　　3. Morals Coordinating Center

d. Parietal　　　4. Hearing

7-68.　　　When you become concerned about the ethical and moral standards of your community, the (1) _frontal_ lobe of your cerebrum is activated.

Hearing is centered in the (2) _temporal_ lobe and sight in the (3) _occipital_ lobe of your cerebral hemispheres.

Your sense of touch is located in the (4) _parietal_ lobe.

7-69.　　　The activities of the cerebral lobes are not as specific as the examples used in this program. Each lobe has additional activities not included here. You have been exposed to the more commonly related _activities_ of each of the cerebral lobes.

7-70.   A MESIAL (MES-i-al)--middle--view of one hemisphere of the brain reveals another lobe, the _cerebellum_, and the brain stem.

NOTE: You should remember that this illustration is representative of but one of the two cerebral hemispheres.

LATERAL VENTRICLE

CORPUS CALLOSUM

CEREBRUM

BRAIN STEM

CEREBELLUM (SER-e-bel-um)

---

cerebellum

7-71.   A human brain has, including the two lobes of the cerebellum, a total of how many lobes.
(a) five lobes
(b) ten lobes
(c) eight lobes

(Select either a, b, or c and move to frame 72)

(b)

7-72.   There are ten lobes or subdivisions of the human brain--
_4_ (number) lobes in each cerebral hemisphere
and the two _cerebellum_ lobes.

---

four

cerebellum

7-73.   The cerebellum, like the cerebral lobes, has a specific function -- that of maintaining coordination and balance. If you are unable to walk in a straight line, the (cerebrum/cerebellum) _cerebellum_ may not be functioning properly.

cerebellum   7-74.

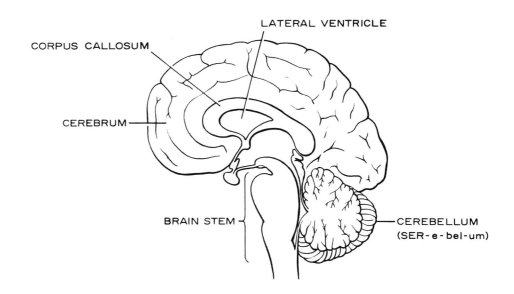

CORPUS CALLOSUM

LATERAL VENTRICLE

CEREBRUM

BRAIN STEM

CEREBELLUM
(SER-e-bel-um)

    An anatomical structure, the <u>CORPUS CALLOSUM</u> (KOR-pus ka-LO-sum), lies <u>superior</u> to the laterial ventricle. This structure joins together the cerebral hemispheres. Locate the corpus callosum in the illustration above.

(No response required–move to frame 75)

7-75.   The corpus callosum is also a communication bridge between the cerebral hemispheres. An impulse from the left side of your brain may be transferred to the right cerebral hemisphere via the ____corpus____ ____callosum____.

---

corpus callosum   7-76.   A mesial surface illustration of the human brain usually contains shaded and non-shaded areas. The non-shaded, or white, areas of the brain are concentrations of <u>nerve fibers</u> called <u>WHITE MATTER</u>. The corpus callosum contains mostly nerve fibers and is therefore a large mass of ____white____ ____matter____.

---

white matter   7-77.   <u>GRAY MATTER</u> is composed of cell bodies; white matter is composed of concentrations of nerve ____fibers____.

fibers	7-78. Nerve cell fibers in the <u>brain</u> and <u>spinal cord</u> are primarily located in the (white/gray) _____*white*_____ matter.
	Nuclei in the brain and spinal cord are primarily located in the (white/gray) _____*gray*_____ matter.
white gray	7-79. Locate and label the gray and white matter in the illustration below.

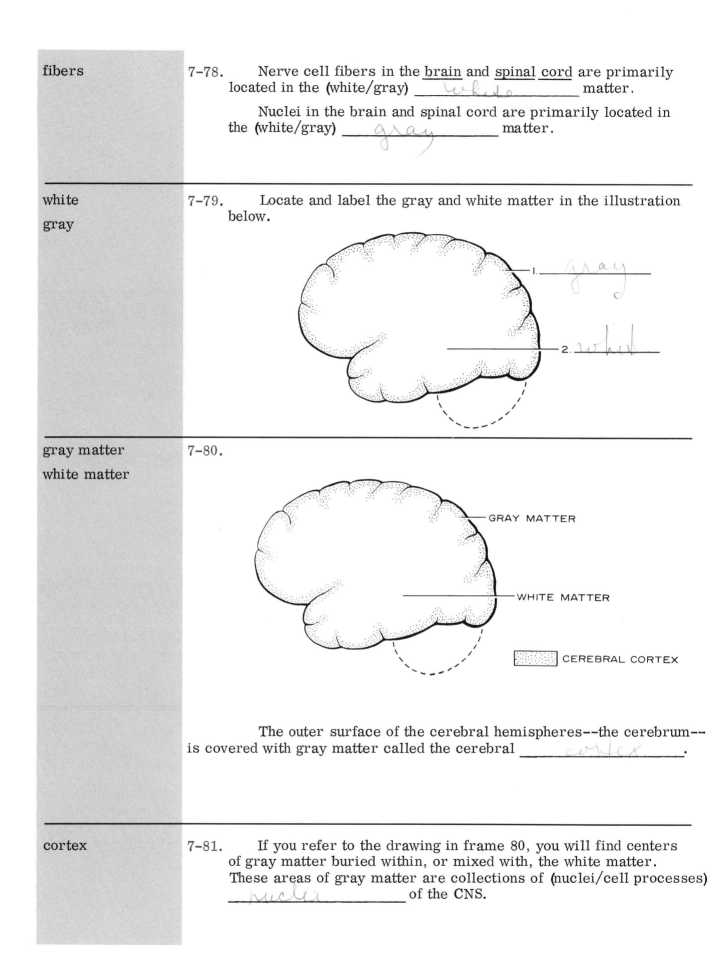

nuclei	7-82. The outer portion of the cerebrum or (1) _cerebral cortex_ is composed primarily of cell bodies or (2) _gray_ matter. A large mass of nerve fibers or (3) _white_ matter, the (4) _corpus callosum_, is one communicating link between the two cerebral hemispheres.
1. cerebral cortex 2. gray 3. white 4. corpus callosum	7-83. There are several bridges of white matter which connect the cerebellum to the BRAIN STEM. These bridges would, like the corpus callosum, be composed of (nerve fibers/cell bodies) _nerve fibers_.
nerve fibers	7-84. The connective bridges, peduncles, are composed of _white_ matter. These peduncles connect the cerebellum to the brain _stem_. SUPERIOR PEDUNCLE (PE-dung-kl) BRAIN STEM— —CEREBELLUM —INFERIOR PEDUNCLE —MIDDLE PEDUNCLE
white stem	7-85. The three bridges that connect the cerebellum to the brain stem are called _peduncles_.
peduncles	7-86. Impulse communication between the <u>cerebellum</u> and <u>brain stem</u> is maintained via the masses of white matter, the (peduncles/ corpus callosum) _peduncles_.

peduncles	7-87. The cervical--neck--region of the spinal cord is connected to the brain proper by the _brain_ _stem_ .

CORPUS CALLOSUM

BRAIN STEM

SPINAL CORD |
| brain stem | 7-88.

PONS (ponz)

MEDULLA OBLONGATA (me-DUL-a ob-long-GA-ta)

SPINAL CORD

The brain stem is composed of two parts, the _pons_ and the _medulla_ _oblongata_ . |
| pons

medulla oblongata | 7-89. The area of the brain stem which is in direct contact with the spinal cord is the (pons/medulla oblongata) _medulla oblongata._ . |
| medulla oblongata | 7-90. The medulla oblongata is the connecting link between the spinal cord and the _pons_ of the brain stem. |

7-91.   BRAIN STEM

Fill in the numbered blanks.

---

1. pons	7-92. The surface of the pons is a mass of white matter; it is there-fore composed of nerve _fibers_ .
2. medulla oblongata	There is gray matter or nerve cell _bodies_ within the pons.

---

fibers	7-93. The medulla oblongata contains nuclei for transmitting impulses from the spinal cord. This section of the brain stem must, therefore, contain (gray/white) _gray_ matter.
bodies	

---

gray	7-94. Sensory impulses are transmitted by (1) _afferent_ nerve fibers to the spinal cord. These impulses then may be communicated to the brain by passing up the spinal cord and through the (2) _medulla oblongata_ and (3) _pons_ of the brain stem.

---

1. afferent	7-95. It is not difficult to determine one function of the brain stem because of its position. The brain stem is the _communications_ "link" between the brain and the spinal cord.
2. medulla oblongata	
3. pons	

communications	7-96. In addition to its function as a conductor of impulses, the medulla oblongata is a reflex center for the circulatory and respiratory systems. Your rate of ___heart___ beat and your rate of ___respiration___ are influenced by the medulla oblongata. ___breathing___
heart breathing	7-97. The pons has two functions: that of conducting impulses from the medulla oblongata to the ___brain___ and that of regulating other body functions.
brain	7-98. Impulses that originate in the nuclei of the medulla oblongata are transmitted to the brain via the ___pons___ .

pons	7-99. There are 12 <u>pairs</u> of <u>CRANIAL</u> (KRA-ni-al) nerves,which originate in the brain and brain stem.
	    Nuclei 1-3 of these ___cranials___ nerves are located in the cerebrum.
	    Nuclei 4-8 of these fibers are located in the pons.

cranial	7-100. Nuclei of cranial nerve pairs 9-10 are located in the medulla oblongata, and the ___11th___ pair of cranial nerves has nuclei in both the <u>medulla oblongata</u> and spinal cord.
11th	7-101. The cranial nerves are arranged in ___pairs___ so that each side of the body can be independently innervated.
pairs	7-102. Efferent impulses that influence circulation and/or respiration, may be transmitted <u>from</u> nuclei of the medulla oblongata, on ___cranial nerve___ pairs 9-11, to the effector organ.

7-103. Superior and anterior to the brain stem, on either side of the 3rd ventricle, are the THALAMUS (THAL-a-mus) and HYPO-THALAMUS (HI-po-THAL-a-mus), areas which are principally masses of gray matter.

HYPOTHALAMUS
CORPUS CALLOSUM
THALAMUS
PONS
MEDULLA OBLONGATA
SPINAL CORD

(No response required—move to frame 104)

7-104. Locate and label the thalamus and hypothalamus. After you have done so, check your labeling with the illustration in frame 103.

thalamus

hypothalamus

7-105. In between the corpus callosum and the brain stem, on either side of the 3rd ventricle, are masses of gray matter, the *thalamus* and the *hypothalamus.*

7-106. A mass of gray matter, the (thalamus/hypothalamus) *thalamus* , is directly below the corpus callosum.

Refer to the illustration in frame 103 if you are unsure of the location of either the thalamus or hypothalamus.

thalamus	**7-107.** Below the thalamus is another mass of gray matter, the *hypothalamus*.
hypothalamus	**7-108.** Although the thalamus and hypothalamus are covered by a thin layer of white matter, they contain a number of cell bodies and therefore consist largely of *gray matter*.

gray matter	**7-109.** The hypothalamus may contain nuclei involved in relaying the impulses of sight and smell.
	    The hypothalamus also contains reflex centers for the regulation of body temperature.
	    The thalamus contains nuclei that transmit afferent *sensory* impulses to the sensory areas of the cerebral cortex.

sensory	**7-110.** Elevation or lowering of the body temperature is influenced by the:
	
	    (a) thalamus (move to frame 112)
	    (b) hypothalamus (move to frame 111)

(b)	**7-111.** Right! The hypothalamus does contain nuclei that influence, not entirely control, the temperature of the body. (No response required—move to frame 113).
(a)	**7-112.** The thalamus does not influence the temperature of the body; this function is one of the activities of the hypothalamus. (No response required—move to frame 113).

	7-113. The thalamus can influence the sleep and appetite control centers.
	    Thalamic functions:
	        1. relay station for (1) *sensory* impulses being transmitted to the cerebral cortex.
	        2. influence on the (2) *sleep* and (3) *appetite* control centers.

1. sensory 2. sleep (either 3. appetite order)	7-114. The thalamus and hypothalamus are located on either side of the 3rd _____ventricle_____, between the corpus callosum and the brain _____stem_____.
ventricle stem	7-115. The two cerebral hemispheres of the human brain are convoluted with deep furrows--_____fissures_____, shallow indentations--_____sulci_____, and rounded elevations -- gyri.
fissures sulci	7-116. Each cerebral hemisphere has _____4_____ (number) main subdivisions or _____lobes_____.
four lobes	7-117. In the illustration below, label the lobes of the cerebral hemisphere, and below each label identify the functional activity of each lobe. FRONTAL (morals) 1. parietal touch 2. temporal hearing 3. occipital sight
1. parietal touch 2. temporal hearing 3. occipital sight	7-118. There are ten main lobes of the human brain: the four lobes found in each of the two cerebral hemispheres and the two lobes of the _____cerebellum_____.
cerebellum	7-119. The cerebellum functions as a coordinating center of: (a) touch (b) sight (c) balance (d) ethics (Select either a, b, c, or d and move to frame 120)

(c)	7-120. A large band of white matter that joins each cerebral hemisphere is the: (a) superior peduncles (b) corpus callosum (Select either a or b and move to frame 121)
(b)	7-121. The corpus callosum may act as an impulse _conductor_ between the two cerebral hemispheres.
conductor (communicator)	7-122. The corpus callosum is composed principally of nerve fibers, which are (gray/white) ____white____ matter.
white	7-123. A group of neuron cell bodies within the CNS is called a ____nucleus____. Such an aggregate of neuron cell bodies is (gray/white) ____gray____ matter.
nucleus gray	7-124. The cerebral cortex is the outer limit of the: (a) cerebrum (b) cerebellum (Select either a or b and move to frame 125)
(a)	7-125. The cerebral cortex is composed primarily of cell bodies, or ____gray____ matter.
gray	7-126. The bridges of white matter that connect the cerebellum with the brain stem are called: (a) corpus callosum (b) peduncles (Select either a or b and move to frame 127)

(b)	7-127. Communication of impulses between the cerebellum and the brain stem travels via the _peduncles_.
peduncles	7-128. The cervical section of the spinal cord is connected to the brain proper by the _brain_ _stem_.
brain stem	7-129. The upper portion of the brain stem is the _pons_ and the lower section is the _medulla_ _oblongata_.
pons medulla oblongata	7-130. Because of the nuclei present within the medulla oblongata, this section of the brain stem contains _gray_ _matter_.
gray matter	7-131 Of the following, what two functions can be related to the functions of the medulla oblongata? (a) impulse communicator (b) sight center (c) touch center (d) reflex center for respiration and circulation
(a) and (d)	7-132. There are _12 pairs_ cranial nerves.

225

7-133.

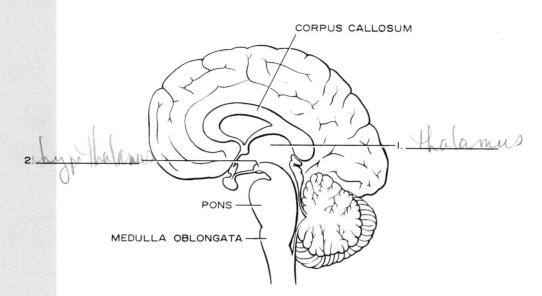

CORPUS CALLOSUM

1. *thalamus*

2. *hypothalamus*

PONS

MEDULLA OBLONGATA

Fill in the unlabeled sections of this drawing.

---

1. thalamus

2. hypo-
   thalamus

7-134.   The two masses of gray matter on either side of the 3rd
ventricle are the ___*thalamus*___ and ___*hypothalamus*___.

---

hypothalamus

thalamus

(either order)

7-135.   BRAIN AREA   FUNCTIONS

(a) Thalamus   1. Sight

(b) Hypothalamus   2. Smell

b -3,

3. Reflex center for the regulation
   of temperature

4. Transmits sensory impulses to
   the cerebral cortex.

4, 5 - a

5. Influences the sleep and appetite
   control centers.

Match the brain areas with the related functions.

(a) 4-5

(b) 1 - 2 - 3

7-136.   This concludes your study of THE NERVOUS SYSTEM –
PART ONE.

Take a break before you begin PART TWO. After your break,
continue into Part Two by turning to the next page.

CHAPTER 7

Part 2 -

The Nervous System

7-137.   Below the medulla oblongata and continuous with the brain stem is the second major part of the CNS, the _spinal_ _cord_ .

---

spinal cord

7-138.   The spinal cord is approximately 18 inches long and 3/4 of an inch in diameter, and because the vertebral column outgrows the spinal cord, the spinal cord in adults (does/does not) _____ extend to the coccyx.

---

does not

7-139.   The spinal cord, like the brain, is protected by bone. The skeletal structure that protects the spinal cord is the _vertebral_ _column_ .

227

7-140.   You will recall, from the discussion of the vertebral column in Chapter 4, that a typical vertebra consists of a rounded *body* anteriorly or ventrally and a *spinous* process at the posterior portion or dorsally.

THORACIC VERTEBRA

DORSAL

SPINOUS PROCESS —

— ARCH

PEDICLE
(PED-i-cul)

BODY

VERTEBRAL FORAMEN —
OCCUPIED BY SPINAL CORD

VENTRAL

7-141.   The PEDICLE (PED-i-cul), which is located between the body and spinous process of the vertebra, has a notch on the inferior (lower) and superior (upper) surface. When the surfaces of this notch are opposite a corresponding notch, in vertebrae either above or below, a foramen (small opening) is formed through which *spinal* nerves pass to the spinal cord.

228

spinal

7-142.

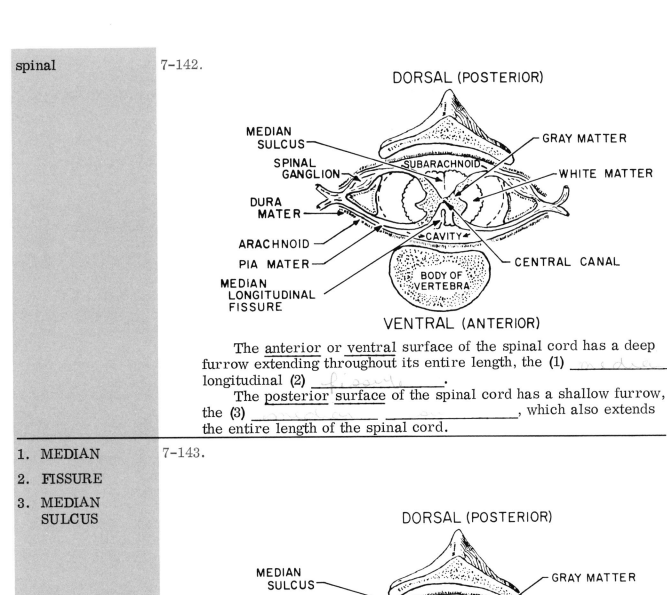

The <u>anterior</u> or <u>ventral</u> surface of the spinal cord has a deep furrow extending throughout its entire length, the (1) _median_ longitudinal (2) _fissure_ .

The <u>posterior surface</u> of the spinal cord has a shallow furrow, the (3) _____ _____ , which also extends the entire length of the spinal cord.

1. MEDIAN

2. FISSURE

3. MEDIAN SULCUS

7-143.

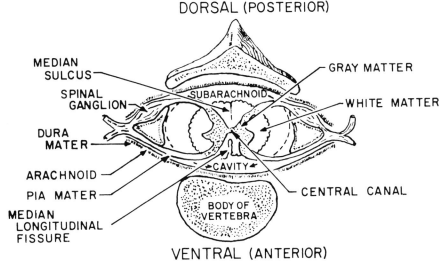

A cross section of the spinal cord reveals (1) _gray_ matter surrounded by (2) _white_ matter. This arrangement of white and gray matter in the spinal cord is (the same as/ opposite from) (3) _____ the arrangement of white and gray matter in the human brain.

1. gray
 	
2. white
 	
3. opposite
   from
	7-144. Refer to the illustration in frame 143.
	
	    This cross section of the spinal cord also reveals that the spinal cord receives protection and support from the (1) _dura_ mater, (2) _pia_ mater, arachnoid and (3) _c. s. f._

1. pia 2. dura (either order) 3. c.s.f.	7-145. The gray matter of the spinal cord conforms roughly to the shape of the letter "H" (refer to the illustration frame 143), with a central cavity (in the crossbar) which contains c.s.f., the _central canal._ .
central canal	7-146. The dorsal and ventral portions of each lateral (arms of the "H") half of the gray matter are commonly called "HORNS." However, since the gray matter extends throughout the entire _length_ of the spinal cord, these areas are also commonly referred to as COLUMNS.
length	7-147. The horns of gray matter or _columns_ of the spinal cord contain, as does all gray matter, _cell bodies_ .
columns cell bodies	7-148. Cell bodies of ventral columns are connected to _efferent motor_ nerve fibers or roots. Cell bodies of dorsal columns are connected to _sensory_ nerve fibers or roots.

DORSAL

CENTRAL GRAY COMMISSURE   SPINOUS PROCESS

DORSAL ROOT GANGLION   SENSORY (AFFERENT) ROOT

IMPULSE   SUBARACHNOID

CAVITY   NERVE TRUNK   _Motor_

BODY OF VERTEBRA   MOTOR (EFFERENT) ROOT

Motor.

VENTRAL

∨ E

ventral – efferent
dorsal – afferent
D A

230

motor (efferent) sensory (afferent)	7-149. A sensory impulse enters the dorsal spinal column via a ganglion known as the _dorsal root ganglion_. This impulse may then: 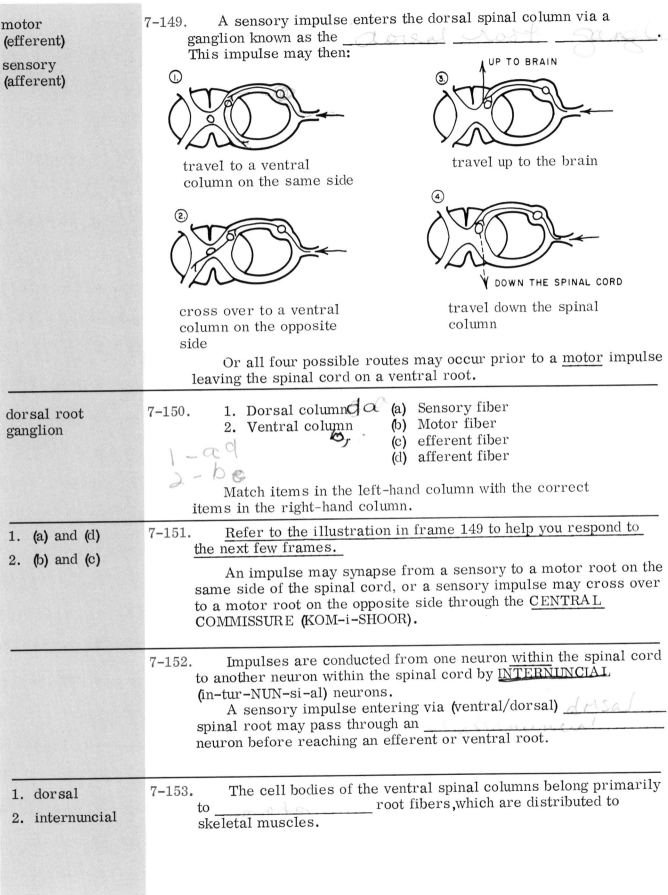 ① travel to a ventral column on the same side ③ travel up to the brain ② cross over to a ventral column on the opposite side ④ travel down the spinal column Or all four possible routes may occur prior to a <u>motor</u> impulse leaving the spinal cord on a ventral root.

dorsal root ganglion	7-150. 1. Dorsal column (a) Sensory fiber
	     2. Ventral column (b) Motor fiber
	                    (c) efferent fiber
	                    (d) afferent fiber
	1—a d
	2—b e
	Match items in the left-hand column with the correct items in the right-hand column.

1. (a) and (d) 2. (b) and (c)	7-151. Refer to the illustration in frame 149 to help you respond to the next few frames. An impulse may synapse from a sensory to a motor root on the same side of the spinal cord, or a sensory impulse may cross over to a motor root on the opposite side through the CENTRAL COMMISSURE (KOM-i-SHOOR).
	7-152. Impulses are conducted from one neuron <u>within</u> the spinal cord to another neuron within the spinal cord by INTERNUNCIAL (in-tur-NUN-si-al) neurons. A sensory impulse entering via (ventral/dorsal) _dorsal_ spinal root may pass through an _internuncial_ neuron before reaching an efferent or ventral root.
1. dorsal 2. internuncial	7-153. The cell bodies of the ventral spinal columns belong primarily to _____ root fibers, which are distributed to skeletal muscles.

motor (efferent)	7-154. DORSAL The cell bodies of all sensory neurons of spinal nerves are lodged in a ___spinal___ ___ganglion___ in a ___dorsal___ (dorsal/ventral) root.
spinal ganglion dorsal	7-155. The spinal cord is a reflex center. Sensory impulses can pass from a dorsal root ganglion via an (1) ___internuncial___ neuron to a ventral (2) ___motor___ root and leave the cord without going to the (3) ___brain___.
1. internuncial 2. motor 3. brain	7-156. Impulses may enter and leave the spinal cord on the same side, or an impulse may cross over to the opposite side of the spinal cord via the central ___gray___ ___commissure___.
gray commissure	7-157. The white matter of the spinal cord, which is located ___around___ (within/around) the gray matter, consists of nerve fibers coursing in all directions but mostly longitudinally.
around	7-158. Impulses which are transmitted to or from the brain travel along the longitudinal TRACTS of the ___white___ matter.

white	7-159. Large collections of spinal cord white matter are referred to as the <u>FUNICULI</u> (fu-NIK-u-li).
	
	        The nerve pathways to or from the brain called _tracts_ are located in the _funiculi_ of the spinal cord.
	
	        NOTE: plural - <u>funiculi</u> -- singular - <u>funiculus</u>

tracts funiculi	7-160. Spinal "horns" or columns refer to the (1) _gray_ (gray/white) matter. The large collections of white matter within the spinal cord, the (2) _funiculi_, contain (3) _tracts_ which carry impulses to and from the brain.
1. gray 2. funiculi 3. tracts	7-161. Tracts that ascend <u>to</u> the brain would be called _sensory_ (sensory/motor) tracts, and tracts that descend from the brain would be called _motor_ (sensory/motor) tracts.
sensory motor	7-162. The <u>SPINOTHALAMIC</u> tract is a sensory tract because it conducts impulses from the _spinal cord_ to the _thalamic_ portion of the brain.
spinal cord thalamic	7-163. The <u>CEREBROSPINAL</u> tract conducts impulses from the brain <u>down</u> the spinal cord. This tract is a (sensory/motor) _motor_ tract.
motor	7-164. There are 31 pairs of spinal nerves arising from the spinal cord. In addition to the spinal nerves, there are 12 pairs of cranial nerves which makes a total of _43 pairs_ (number) cranio-spinal nerves.
86 nerves or 43 pairs	7-165. Cranial nerves <u>are</u> similar <u>functionally</u> to the spinal nerves. Cranial nerves <u>are not</u> exactly similar to the spinal nerves structurally. You (would/would not) _would not_ expect all the _31_ (number) pairs of spinal nerves to be anatomically like the cranial nerves.

would not 31	7-166 You should keep in mind the definition of a nerve. You will recall that a bundle of nerve _*fibers*_ constitutes a nerve.
fibers	7-167 In some nerves there <u>are</u> both sensory and _*motor*_ fibers that carry impulses to or away from the <u>CNS</u>.

7-168

    Spinal nerves and some cranial nerves contain both _*motor*_ and _*sensory*_ fibers.

motor (efferent)

motor sensory (either order)	7-169 Spinal <u>roots</u> (are/are not) _*are not*_ composed of mixed fibers. Where an efferent and afferent root unite, a _*spinal nerve*_ is formed.
are not spinal nerve	7-170 Fibers from various spinal nerves, after leaving the spinal cord, intermingle and form networks of nerves. This <u>intermingling</u> of <u>nerves</u> and <u>fibers</u> is called a <u>PLEXUS</u> (PLEK-sus). Spinal nerve fibers from one <u>level</u> in the neck intermingle with <u>other</u> spinal <u>nerve</u> fibers from <u>different</u> levels in the neck to form the <u>CERVICAL</u> _*plexus*_.

plexus	7-171	A plexus is an intermingling of (1) _____ and (2) _____ .
		At the lower end of the spinal cord, spinal nerves leave the cord and intermingle with nerves and fibers from other levels of the spinal cord to form the LUMBOSACRAL (3) _plexus_.
1. nerves 2. fibers 3. plexus	7-172	Such an intermingling of nerves and fibers allows for a diffuse reaction. For example, if you step on a nail, the impulse returning to the spinal cord on sensory fibers reaches the cord where the impulse can be conducted to many levels of the spinal cord on several of the intermingling sensory fibers, <u>outside</u> the cord, in the lumbosacral plexus.
		You react by lifting your foot, but you may also _____ (use your own words to describe additional reactions), because the original stimulus has been conducted to many areas of the spinal cord and brain.
Yell, dance about, hold your foot (Use your own words)	7-173	The PERIPHERAL (per-RIF-er-al) NERVOUS SYSTEM includes the nerves which connect the body parts to the CNS. This system would include the 43 pairs of _cranial_ and _spinal_ nerves.
cranial spinal (either order)	7-174	Craniospinal nerves that innervate the periphery of the body constitute the _peripheral_ nervous system.
peripheral	7-175	The fibers that compose the peripheral nervous system are: (a) – sensory and motor fibers (b) – mostly sensory fibers (c) – mostly motor fibers (Select either a, b, or c and then move to frame 176).
(a)	7-176	Peripheral nerves contain both sensory and motor fibers because they must conduct impulses from the receptor organs to the spinal cord and from the spinal cord via _peripheral motor_ fibers to effector organs, usually skeletal muscle.

7-177

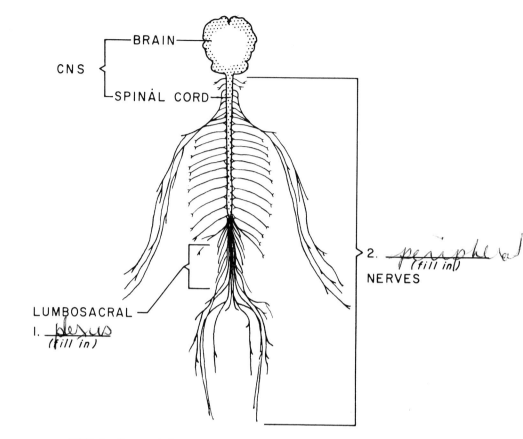

BRAIN

CNS

SPINAL CORD

2. *peripheral*
   (fill in)
   NERVES

LUMBOSACRAL —

1. *plexus*
   (fill in)

Fill in the unlabeled blanks.

---

1. plexus
2. peripheral

7-178   Your spinal cord:

(a) extends the entire length of your vertebral column.
(b) does not extend the entire length of your vertebral column.

(Select either a or b, and then move to frame 179).

(b)

7-179

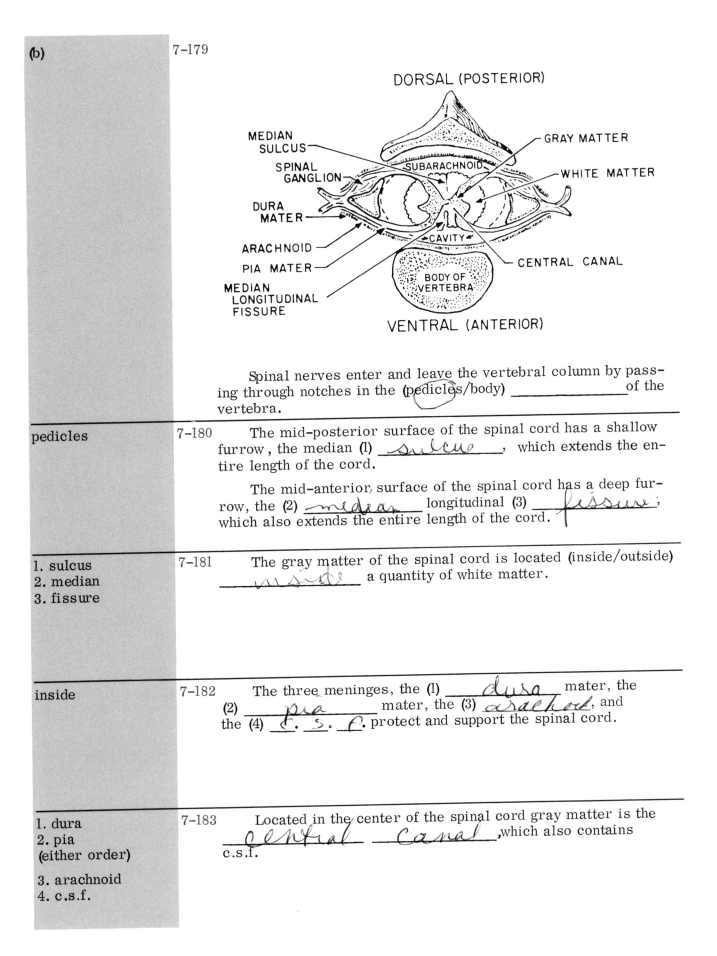

DORSAL (POSTERIOR)

MEDIAN SULCUS

SPINAL GANGLION

DURA MATER

ARACHNOID

PIA MATER

MEDIAN LONGITUDINAL FISSURE

SUBARACHNOID

CAVITY

BODY OF VERTEBRA

GRAY MATTER

WHITE MATTER

CENTRAL CANAL

VENTRAL (ANTERIOR)

Spinal nerves enter and leave the vertebral column by passing through notches in the (pedicles/body) _____ of the vertebra.

---

pedicles

7-180   The mid-posterior surface of the spinal cord has a shallow furrow, the median (1) __sulcus__, which extends the entire length of the cord.

The mid-anterior surface of the spinal cord has a deep furrow, the (2) __median__ longitudinal (3) __fissure__; which also extends the entire length of the cord.

---

1. sulcus
2. median
3. fissure

7-181   The gray matter of the spinal cord is located (inside/outside) __inside__ a quantity of white matter.

---

inside

7-182   The three meninges, the (1) __dura__ mater, the (2) __pia__ mater, the (3) __arachnoid__, and the (4) __c. s. f.__ protect and support the spinal cord.

---

1. dura
2. pia
(either order)

3. arachnoid
4. c.s.f.

7-183   Located in the center of the spinal cord gray matter is the __central canal__, which also contains c.s.f.

central canal	7-184	The dorsal and ventral portions of the gray component of the spinal cord are called: (a) "horns" (b) tracts (c) columns (Select either a, b, or c, and then move to frame 185).
(a) or (c)	7-185	The dorsal spinal columns contain the (motor/sensory) (1) _sensory_ cell bodies. Fibers of these cell bodies synapse with internuncial neurons, which in turn (2) _synapse_ with cell bodies of the (3) _motor_ root. Ventral spinal columns contain cell bodies of the (motor/sensory) (4) _____ root.
1. sensory 2. synapse 3. motor (ventral) 4. motor	7-186	If you are slapped on the back, the sensory impulse passes via a spinal nerve to a (1) _dorsal_ root, through a dorsal root (2) _ganglion_ to the dorsal spinal columns. The impulse is then conducted to the brain or via an (3) _internuncial_ neuron to the (4) _ventral_ motor root.
1. dorsal 2. ganglion 3. internuncial 4. ventral	7-187	Impulses that enter the spinal cord can pass, via an internuncial neuron to: (a) efferent fibers on the same side of the cord (b) efferent fibers on the opposite side of the cord (c) both of the above (Select either a, b, or c, and then move to frame 188).
(c)	7-188	A sensory impulse crossing over from a dorsal column to the ventral column of the opposite side of the cord must pass through the central gray _____ (commissure/anterior root).
commissure	7-189	Spinal _tracts_ , which carry impulses to or from the brain, are located in the _white_ (gray/white) matter.
tracts white	7-190	The fibrous tracts are located in the large areas of the spinal cord white matter the: (a) arachnoid cavity (b) dorsal root (c) funiculi (Select either a, b, or c, and then move to frame 191).

(c)

7-191   A cerebrospinal tract is a:

(a) sensory tract to the brain
(b) motor tract from the brain

(Select either a or b, and then move to frame 192).

(b)

7-192   There are _____12_____ (number) pairs of cranial nerves
and ____31____ (number) pairs of spinal nerves.

12

31

7-193   Spinal nerves and some (1) ____cranial____ nerves contain both (2) ____sensory____ and (3) ____motor____ fibers.

---

1. cranial

2. motor (either

3. sensory order)

7-194   A network of intermingling nerves and fibers outside the
spinal cord is called a:

(a) root
(b) plexus

(Select either a or b and then move to frame 195).

(b)

7-195   An impulse travels from the CNS via a nerve plexus to the
periphery of the body inducing a reaction in the ____effector____
organs usually made up of (skeletal/smooth) ____skeletal____
muscle.

---

effector
skeletal

7-196   The peripheral nervous system consists of the 12 pairs of
____cranial____ and the 31 pairs of ____spinal____
nerves.

---

cranial
spinal

7-197   Because the (1) ____peripheral____ nervous system must
conduct impulses to and from the periphery of the body, the nerves
of this system must contain (2) ____motor____ and (3) ____sensory____
fibers and be functionally integrated with other nervous tissue of
the body.

1. peripheral 2. sensory (either 3. motor order)	7-198	Take a break before you continue.

<table>
<tr><td></td><td>7-199</td><td>A group of neuron cell bodies <u>outside</u> the CNS is called a (nucleus/ganglion) <u>ganglion</u> .</td></tr>
</table>

---

ganglion	7-200	The cell bodies of dorsal root fibers make up the dorsal root <u>ganglion</u> . Dorsal root fibers synapse with cell bodies <u>within</u> the dorsal columns of the spinal cord. These cell bodies of the dorsal columns are referred to as a (nucleus/ganglion) <u>nucleus</u> .

---

ganglion nucleus	7-201	Impulses leave the spinal cord via (dorsal/ventral) (1) <u>ventral</u> roots and travel via the (2) <u>peripheral</u> nervous system to the (3) <u>effector</u> organs.

---

1. ventral 2. peripheral 3. effector	7-202	From our earlier discussion of the nervous system, you found that although the nervous system is generally divided into three anatomically separate systems: The Central Nervous System (CNS) The Peripheral Nervous System The Autonomic (0-to-NOM-ik) Nervous System (ANS) the nervous tissue of the body is <u>functionally</u> integrated.

---

functionally	7-203	There is a lack of uniformity in the terminology which is used to describe the <u>Autonomic Nervous System</u> (ANS). It is referred to as: The Visceral Nervous System, because of its involvement with the internal organs (viscera); <u>Vegetative Nervous</u> System, because of its involuntary or unconscious functions; or simply <u>The Involuntary Nervous System</u>. (No response required, move to frame 204).

7-204.   The term, Autonomic Nervous System, indicates that this system is autonomous or that it functions independently of the CNS and peripheral nervous systems, however; this is not so. The ANS is _functionally integrated_ with other nervous tissue of the body.

---

functionally

integrated

7-205.   The system that controls the functions of the viscera (organs), which are generally not under voluntary control is the _A N S_.

ANS
(Autonomic Nervous System)

7-206.   Voluntary actions - movement of skeletal muscles - are regulated by the (1) _central_ and (2) _peripheral_ nervous systems.

Involuntary actions - contractions or relaxations of smooth and cardiac muscles and secretions of many glands - are regulated by the (3) _ANS._ nervous system.

---

1. central

2. peripheral
(either order)

3. autonomic

7-207.   There are two divisions of the ANS with distinct connections to the CNS: the SYMPATHETIC (sim-pa-THET-ik) and the PARASYMPATHETIC (par-a-sim-pa-THET-ik).

The two divisions of the CNS are the brain and spinal cord. The two divisions of the ANS are the _Sympathetic_ and _parasympathetic_

---

sympathetic

parasympathetic

(either order)

7-208.   Fill in the four unlabled blanks

SYSTEM	DIVISIONS
CNS	1. _central_ brain 2. _peripheral_ spinal cord
ANS	3. _sympa._ 4. _parasympa._

1. Brain	**7-209.** The sympathetic <u>preganglionic</u> fibers originate from cell bodies in the lateral columns of the spinal cord gray matter, from approximately the level of the eighth <u>cervical</u> to the third lumbar vertebrae. This division of the ANS is, therefore, called the:
2. spinal cord	
(either order)	
3. sympathetic	(a) Thoracicolumbar outflow
4. parasympathetic	(b) Craniosacral outflow
(either order)	(Select either a, or b, and then move to frame 210).

(a)   **7-210.** The sympathetic system of the ANS is called the THORACICOLUMBAR (tho-RAS-i-ko-LUM-ber) outflow because it is from these areas of the vertebral column that the axons of the sympathetic system leave the *Spinal cord*.

---

spinal cord   **7-211.** After the preganglionic fibers of the sympathetic neurons leave the spinal cord, they mingle within fibers of the ventral roots of the spinal nerves and may then *synapse* with autonomic ganglia outside the CNS.

---

synapse   **7-212.**

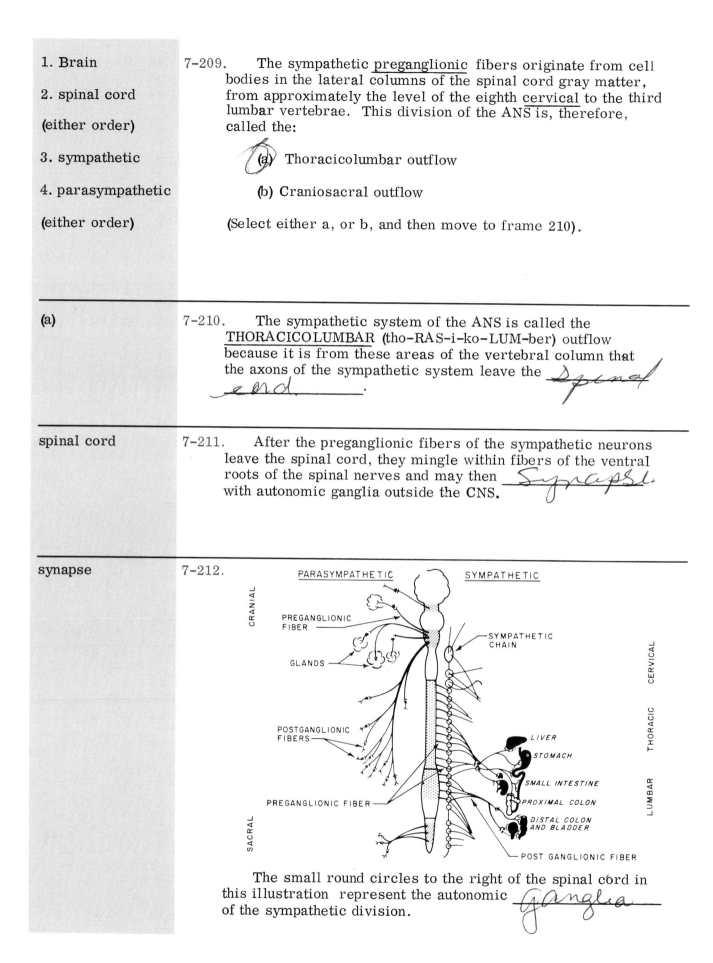

The small round circles to the right of the spinal cord in this illustration represent the autonomic *ganglia* of the sympathetic division.

ganglia	7-213. The ANS fibers that connect the ganglia to the spinal column are called:

(a) preganglionic fibers
(b) postganglionic fibers

(Select either a, or b, and then move to frame 214). |
| (a) | 7-214.   Preganglionic fibers of the ANS are myelinated and are referred to as the _____white_____ (white/gray) RAMI (RA-mi) (the plural of Ramus) COMMUNICANTES (co-mu-ni-CAN-tes). |
| white | 7-215.   The preganglionic fibers or white _____rami_____ _____communicantes_____ are insulated with a white, fatty, noncellular material known as _____myelin_____. |
| rami communicantes

myelin | 7-216.   There are three routes which the sympathetic preganglionic fibers can follow.

After leaving the spinal column via the _____white rami_____ communicantes, the fibers can separate from the _____peripheral_____ nerves and synapse immediately with the neurons of the nearest ganglion in the sympathetic division. |
| white rami

peripheral | 7-217.

In the above drawing, you can see one set of 22 sympathetic ganglia, which are interconnected to form a chain. There are two sympathetic chains, each with 22 _____sympathetic_____ _____ganglia_____, one chain on each side of the vertebral column. |

sympathetic ganglia	7-218. A second route of impulse transmission for the sympathetic preganglionic fibers would be to enter the sympathetic chain and travel <u>up</u> or <u>down</u> the chain _synapsing_ with cells in distant ganglia of the chain.
synapsing	7-219. Finally the white _rami communicantes_ can pass through the sympathetic ganglia, without <u>synapsing</u>, leave the sympathetic chain and travel on the _peripheral_ nerves to the periphery of the body where they then synapse with peripheral ganglion cells (cardiac ganglion, superior mesenteric ganglion, etc).
rami communicantes peripheral	7-220. Impulses may be transmitted by the sympathetic preganglionic fibers by three routes. First, an impulse may leave the spinal cord via the (white/gray) _white_ rami communicantes, separate from the peripheral nerves and synapse (immediately/later) _____ with the neurons of a ganglion of the sympathetic division.
white immediately	7-221. Second, because the _22_ (number) sympathetic ganglia on each side of the vertebral column are interconnected, an impulse may be transmitted on preganglionic fibers that travel up or down the sympathetic chain (with/<u>without</u>) _____ synapsing until arriving at a distant ganglion.
22 without	7-222. Third, an impulse can pass on a preganglionic fiber, white _rami_ communicantes, without synapsing through a sympathetic chain neuron, via (autonomic/peripheral) _____ nerves, to distant peripheral ganglia of the body where they then synapse.
rami peripheral	7-223. The preganglionic fibers of the sympathetic division are: (1) (afferent/efferent) _efferent_ in reference to CNS, (2) (afferent/efferent) _afferent_ in regard to the sympathetic ganglia.
(1) efferent (2) afferent	7-224. It is important to remember that the <u>afferent</u> preganglionic fibers of the sympathetic division arise from the lateral columns in the spinal cord. (Few/None) _None_ of the preganglionic fibers arise from sympathetic ganglia.

None	7-225. Preganglionic fibers of the ANS are also called _white rami communicantes_
white rami communicantes	7-226. Preganglionic fibers of the ANS sympathetic division arise from the lateral columns of the _spinal column_ and then synapse with a sympathetic ganglia and are therefore _____ (afferent/efferent) in function.
spinal column afferent	7-227. Fibers that conduct impulses from a sympathetic ganglion to the effector organ are called: (a) preganglionic white rami (b) postganglionic fibers (Select either a or b and then move to frame 228).
(b)	7-228. The afferent preganglionic fibers of the ANS arise from the (1) _lateral columns_ of the spinal cord. Most fibers arising from the sympathetic ganglia (axons of the sympathetic ganglia) are (2) _motor_ (motor/sensory) in function and are called (3) _postganglion_ fibers.
1. lateral columns 2. motor 3. post-ganglionic	7-229. A good example of the functional integration of the ANS with other nervous tissue is a VISCERAL REFLEX ARC. The peripheral nervous system sensory fibers from an innervated organ conduct an impulse to the _lateral column_ of the spinal cord.
lateral columns	7-230. The impulse then travels from the lateral columns on a _preganglionic_ fiber-white ramus communicantes - which synapses with a _post ganglionic fiber._
preganglionic postganglionic fibers	7-231. The axons of neurons in the sympathetic ganglion which are (pre/post) _post_ ganglionic fibers, carry the impulse back to the innervated organ completing the formation of the Visceral _Reflex arc._

post Reflex Arc	7-232. A myelinated ANS fiber that arises in the lateral columns of the spinal cord is a _pregangleon_ fiber. An unmyelinated ANS fibers that is a continuation of a sympathetic ganglia cell is a _post g_ fiber.
pre- ganglionic post- ganglionic	7-233. Preganglionic sympathetic fibers are _myelinated_ (myelinated/unmyelinated). Postganglionic sympathetic fibers are _unmyelinat_ (myelinated/unmyelinated).
myelinated unmyelinated	7-234. The second division of the ANS, the _parasym_ division, is sometimes called _craniosac_ (thoracicolumbar/craniosacral) outflow.
para sympathetic cranio- sacral	7-235. Refer to the drawing in frame 234. The ganglia of the _parasympa_ division are located close to the spinal cord, and the ganglia of the _sympa_ division are located in or near the organ innervated by the efferent fibers.
sympathetic parasympathetic	7-236. The preganglionic fibers of the parasympathetic division have their origin in the nuclei of the _brain_ and the _sacral_ region of the spinal cord.

brain sacral	7-237. Preganglionic parasympathetic fibers carry impulses to or near the innervated _____*organ*_____. The preganglionic fibers then synapse with a parasympathetic _____*ganglion*_____.
organ ganglion	7-238. Sympathetic ganglia are located (close to/<u>far away from</u>) _____ the innervated organ. Parasympathetic ganglia are located (close to/far away from) _____ the innervated organ.
far away from close to	7-239. Sympathetic postganglionic fibers "fan out" in many directions before they reach the innervated organs. You would therefore expect a wide spread or diffuse body reaction from impulses conducted by the (sympathetic/parasympathetic) _____ division of the ANS.
sympathetic	7-240. The preganglionic fibers of the (1) _____*parasym*_____ division synapse with a ganglion that is (2) _____*close*_____ to or within the innervated organ. The postganglionic fibers of this division are short and more direct. Stimulation of this division results in a (diffuse/<u>local</u>) (3) _____ effect.
1. para- sympathetic 2. close 3. local	7-241. The scope of the ANS functions can be better appreciated if you consider the wide distribution of <u>smooth</u> <u>muscle</u> in blood vessels from head to foot, digestive organs from the mouth to the anus, urinary tract organs, reproductive organs, special sensory organs such as the eye, and with hairs and glands on the body surface. (No response required, move to frame 242)
	7-242. The over-all function of the ANS is to maintain a constant environment for the cells of the body. This includes constant body temperature, regulation of the heart rate and respiratory rate, digestive functions, <u>all</u> the processes vital to cell integrity, and a constant fluid _____*environment*_____.
environment	7-243. In order to live in outer space, an astronaut must be provided with an environment that closely approximates conditions on earth. The atmosphere within the space capsule is such an environment. Your body, like the astronaut's, must maintain a constant environmental state. The _____*ANS*_____ nervous system helps your body when internal and external environmental changes occur.

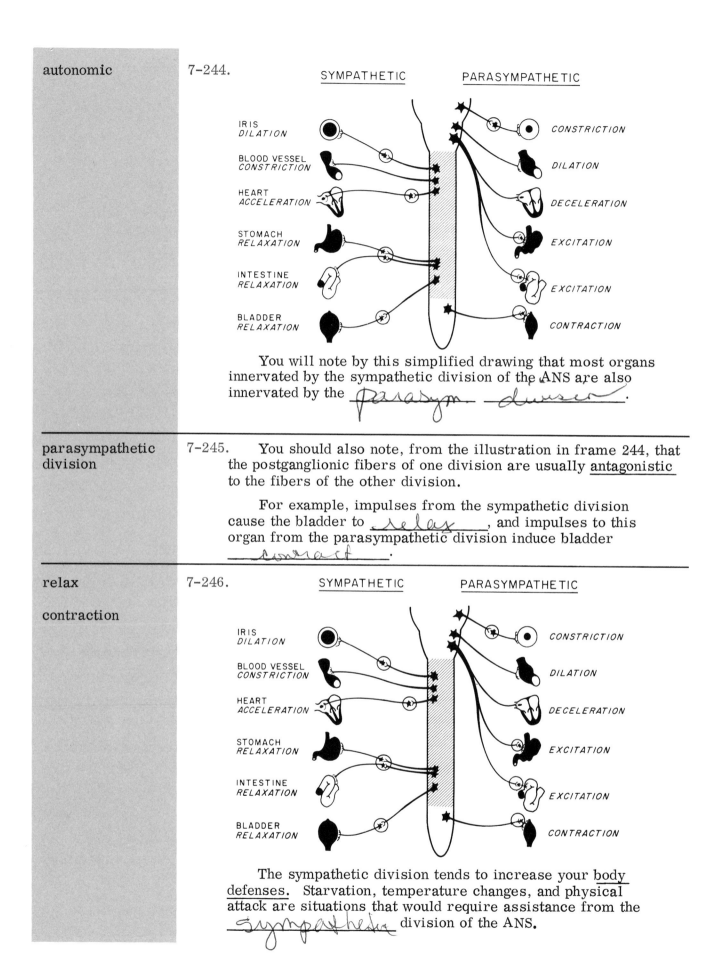

autonomic	**7-244.**

SYMPATHETIC   PARASYMPATHETIC

IRIS
DILATION — CONSTRICTION

BLOOD VESSEL
CONSTRICTION — DILATION

HEART
ACCELERATION — DECELERATION

STOMACH
RELAXATION — EXCITATION

INTESTINE
RELAXATION — EXCITATION

BLADDER
RELAXATION — CONTRACTION

You will note by this simplified drawing that most organs innervated by the sympathetic division of the ANS are also innervated by the *parasym. division*.

---

parasympathetic division	**7-245.** You should also note, from the illustration in frame 244, that the postganglionic fibers of one division are usually <u>antagonistic</u> to the fibers of the other division.

For example, impulses from the sympathetic division cause the bladder to *relax*, and impulses to this organ from the parasympathetic division induce bladder *contract*.

---

relax contraction	**7-246.**

SYMPATHETIC   PARASYMPATHETIC

IRIS
DILATION — CONSTRICTION

BLOOD VESSEL
CONSTRICTION — DILATION

HEART
ACCELERATION — DECELERATION

STOMACH
RELAXATION — EXCITATION

INTESTINE
RELAXATION — EXCITATION

BLADDER
RELAXATION — CONTRACTION

The sympathetic division tends to increase your <u>body defenses</u>. Starvation, temperature changes, and physical attack are situations that would require assistance from the *sympathetic* division of the ANS.

sympathetic	7-247. The parasympathetic division is directed toward the conservation and restoration of the body's resources.

When you become frightened and your heart begins to beat excessively, it is the _sympa._ division of the ANS that restores it to a more normal pace. |
| parasympathetic | 7-248. You are not expected to remember all the functions of both divisions of the ANS; however, you should remember that when your body is called upon to fight and your heart beats faster to provide more energy to the cells, it is the _symp._ division which induces this increased _defense_ action. |
| sympathetic

defensive | 7-249. After your body has met the challenge, it is the parasympathetic division that induces the _restoration_ of used body energy and the _conservation_ of the body's resources in preparation for other emergencies. |
| restoration

conservation | 7-250. The 12 pairs of cranial nerves and 31 pairs of spinal nerves constitute the _peripheral_ nervous system. |
peripheral	7-251. Voluntary actions, such as movement of _skeletal_ muscles, are controlled by the CNS.
skeletal	7-252. Involuntary actions (excluding reflexes of skeletal muscles), such as movement of smooth and cardiac muscles, are regulated by the _ANS_ nervous system.
ANS	7-253. The ANS _is_ (is/is not) functionally integrated with other nerve tissue.

is	7-254. The ANS is divided into two main divisions the thoracicolumbar outflow or _Sym._ division and the craniosacral outflow or _para_ division.
sympathetic parasympathetic	7-255. Sympathetic preganglionic fibers originate from cell bodies in the _lateral columns_ of the spinal cord _gray_ matter.
lateral columns gray	7-256. Another name for the sympathetic division is the thoracicolumbar outflow. This name refers to origin of the sympathetic preganglionic fibers, the (1) _thoracic_, (2) _cervical_, and (3) _lumbar_ areas of the spinal cord.
1. cervical 2. thoracic 3. lumbar	7-257. The ganglia of the sympathetic division of the ANS are located close to the _vertebral column_ (vetebral column/innervated organ).
vertebral column	7-258. Sympathetic preganglionic fibers are _myelinated_ (myelinated/unmyelinated). These fibers are called _white rami communicantes_
myelinated white rami communicantes	7-259. Sympathetic preganglionic fibers can, after leaving the spinal cord: 1. separate from the spinal nerves and immediately synapse with one of the 22 _sym. ganglia_ on each side of the vertebral column,
sympathetic ganglia	7-260. 2. travel up or down the chain of interconnected sympathetic ganglia and synapse with distant _ganglia_ of the sympathetic chain,

ganglia	7-261.　　3. pass without synapsing with a ___Sym___ ___ane___ and be conducted via ___peripheral___ nerves to the effector organ.
sympathetic ganglion peripheral	7-262.　　Sympathetic preganglionic fibers that are afferent in function arise from the <u>lateral columns</u> of the spinal cord and then synapse with a ___sym.___ ___ganglion.___
sympathetic ganglion	7-263.　　All fibers <u>arising from</u> the sympathetic ganglia are (1) ___motor___ in function and are called (postganglionic/preganglionic) (2) ___post.___ fibers. 　　Because these fibers are <u>unmyelinated,</u> they are called (white/gray) (3) ___gray___ · rami communicantes.
1. motor (efferent) 2. postganglionic 3. gray	7-264.　　A visceral reflex arc is formed when the sensory fibers of the ___periphera___ nervous system transmit an impulse from an innervated organ to the ___lateral___ columns of the spinal cord.
peripheral lateral	7-265.　　The impulse is then conducted via (white/gray) ___white___ rami communicantes <u>to</u> a sympathetic ___ganglion___
white ganglion	7-266.　　From the sympathetic chain, the impulse is conducted <u>away</u> from the sympathetic ganglia on (dendrites/axons) (1) ___axons___ of the sympathetic ganglion cells, (white/gray) (2) ___gray___ rami communicantes, or on (3) ___per.___ nerves to the innervated organ.
1. axons 2. gray 3. peripheral	7-267. ·　　A visceral reflex arc is a good example of the ___functional___ ___integrity___ of <u>all</u> nervous tissues.

functional integration	7-268. Preganglionic fibers of the parasympathetic division of the ANS – craniosacral outflow – arise from the gray matter of the _brain_ or sacral region of the _spinal cord_.
brain spinal cord	7-269. Preganglionic parasympathetic fibers of the ANS conduct impulses to or near the innervated organ where the fibers synapse with a _ganglion_.
ganglion	7-270. The ganglia of the _para_ (sympathetic/ parasympathetic) division are located close to or within the innervated organ.
parasympathetic	7-271. Before reaching the innervated organ, an impulse conducted on a _sympathetic_ ANS fiber spreads out in many directions.
sympathetic	7-272. Impulses conducted on _para_ ANS fibers would induce a local effect.
parasympathetic	7-273. The ANS acts to maintain a constant body (1) _environ_ when affected by either (2) _internal_ or (3) _external_ body changes.
1. environment 2. internal (either 3. external order)	7-274. When the sympathetic and parasympathetic divisions of the ANS respond to internal or external body changes, they usually act (synergistically/antagonistically) _antag_.

252

antagonis- tically	7-275. Your body defense system would be activated by the ___*sympa*___ division of the ANS if you are physically attacked.
sympathetic	7-276. The parasympathetic division of the ANS helps your body ___*restore*___ and ___*conserve*___ its resources.
restore conserve (either order)	7-277. Your heart rate can be accelerated by the ___*sym.*___ division of the ANS and decelerated by the ___*para.*___ division of the ANS.
sympathetic parasympathetic	7-278. You may want to take a break before you begin the concluding section of this unit.
	7-279. Your nervous system is, in an operational sense, like a high-speed computer. Data - impulses - are fed into the computer and are then analyzed, and a response, "feedback," is provided. (No response required; move to frame 280).
	7-280. The cerebral cortex, the highest governing level of the nervous system, is, however, much more complex than our most sophisticated computers. You will recall that the cerebral cortex is connected to the brain stem and spinal cord by ascending and descending ___*tracts.*___
tracts	7-281. Some impulses travel directly from the ___*spinal*___ ___*cord.*___ to the cerebral cortex on direct high speed ___*tracts*___.

spinal cord tracts	7-282. There is another connective network between the spinal cord and cerebral cortex. This network consists of small fibers in the spinal cord and brain stem and is called the RETICULAR (re-TIK-u-ler) SYSTEM. An indirect tract to the cerebral cortex would pass through the ___retic___ system.
reticular	7-283. Impulses that pass through the indirect route of small ___retic___ fibers are carried to the cerebral cortex on thalamic fibers.
reticular	7-284. If a sensation is to be experienced, you must be awake. The reticular system keeps the cerebral cortex awake and is therefore referred to as the ___reticular___ activating system.
reticular	7-285. Your cerebral cortex is kept alert to conscious sensations by impulses which originate in the reticular ___activating___ system and are transmitted to the cortex via ___thalamic___ fibers.
activating thalamic	7-286. Impulses enter the CNS on (sensory/motor) (1) ___sensory___ neurons. They pass from the spinal cord directly to the brain on highspeed (2) ___tracts___ or through the small fibers of the reticular activating system, which is located in the spinal cord and (3) ___brain stem___.
1. sensory 2. tracts 3. brain stem	7-287. To experience a sensation, your cerebral cortex must be (1) ___awake___. The reticular (2) ___activating___ system transmits impulses via (3) ___thalamic___ fibers to keep the cortex alerted to external stimuli.
1. awake 2. activating 3. thalamic	7-288. PERIPHERAL MOTOR ACTIVITIES are controlled by the cerebral cortex and nuclei in the brain stem. Actions such as walking and talking are examples of ___peripheral___ motor activities.

254

peripheral	7-289. The <u>PERIPHERAL AUTONOMIC MOTOR ACTIVITY,</u> like other <u>peripheral</u> <u>motor</u> <u>activities</u>, is also controlled by nuclei in the upper _brain_ _stem_ and other sections of the brain.
brain stem	7-290. BRAIN SURFACE-ONE HEMISPHERE MEDIAL VIEW The ring of gray matter in the above drawing is referred to as the autonomic or visceral brain. This area of the brain would control the peripheral _autonomic_ motor activities.
autonomic	7-291. Refer to the illustration in frame 290. The LIMBIC (LIM-bik) area of the ANS is composed of two main sections, the (1) _amygdala_ and the (2) _hippocampus_. The limbic area controls the peripheral (3) _autonomic motor_ activities of the body.
1. amygdala 2. hippocampus (either order) 3. autonomic motor	7-292. The limbic area, consisting of the _amygdala_ and _hippocampus_, is referred to as the <u>limbic system.</u> This system influences the <u>emotional</u> and <u>intellectual</u> aspects of <u>conscious</u> behavior.
amygdala hippocampus (either order)	7-293. Sexual activity – reproduction – and securing of food – survival – are activities which have the most profound influence on your behavior. These activities are under the control of the _limbic_ area of the brain.

255

limbic	**7-294.** The amygdala influences our survival patterns. The hippocampus influences our sexual activities. Match the areas of the brain in the column at the left with the activities in the right-hand column:
	
	    (a) amygdala (1) sexual activities
	    (b) hippocampus (2) emotion and intellect
	    (c) limbic system (3) survival

(a) - 3 (b) - 1 (c) - 2	**7-295.** Peripheral motor activity is controlled by the cerebral cortex of the brain and the _brain stem_. Peripheral autonomic motor activity is controlled by the visceral brain -- _limbic system_ -- and the hypothalamus.
brain stem limbic area	**7-296.** Peripheral autonomic motor activity is regulated and correlated by two sections of the brain, the _hypothalamus_ and the _limbic_ areas.
hypothalamus limbic	**7-297.** It is interesting to note that efferent impulses involving either survival or sexual activities leave the limbic system and travel to peripheral motor and peripheral autonomic motor centers resulting in behavior patterns which involve both the _autonomic_ and _central_ nervous systems.
central autonomic (either order)	**7-298.** The cerebral cortex is kept awake by impulses originating in the _reticular activating_ system. These impulses are then transmitted to the cortex via _thalamic_ fibers.
reticular activating thalamic	**7-299.** The ANS peripheral motor activities are regulated and correlated by the _hypothalamus_ and _limbic_ areas of the brain.
limbic hypothalamus	**7-300.** Another example of the functional integration of nerve tissues is the manner in which afferent impulses involving sexual activity or survival are transmitted from the limbic area on peripheral motor and peripheral autonomic motor nerves of both the _central_ and _autonomic_ nervous systems.

central

autonomic

7-301. Match the brain structures at the left with the functions listed at the right:

STRUCTURE	FUNCTION
(a) Limbic system	1. Conscious thinking
(b) Thalamus	2. Alerts cerebral cortex
(c) Amygdala	3. Correlates emotion and intellect.
(d) Hippocampus	4. Sexual activities
(e) Cerebral cortex	5. Survival
(f) Reticular activating System	6. Transmits alerting impulses

a — 3
b — 6
c — 5
d — 4
e — 1
f — 2

a - 3
b - 6
c - 5
d - 4
e - 1
f - 2

7-302. Impulses leave the brain and brain stem via direct high speed tracts or through the reticular system and pass to either the peripheral motor or peripheral __autonomic__ motor fibers.

autonomic

7-303. You know that impulse transmission along the axon and dendrite of a neuron is a (n) (electrical/mechanical) __electrical__ phenomenon.

electrical

7-304. In a resting state-no impulse transmission, nerve fibers are POLARIZED (PO-ler-iz-ed). A polarized fiber has positive ions (charged atoms) located on the outer surface of the membrane that surrounds the fiber and an equal number of __negative__ ions located on the inner surface of the nerve fiber membrane.

negative

7-305. Nerve fibers are surrounded by a semi-permeable membrane. When an impulse is transmitted along a nerve fiber, the permeability of the fiber is increased, and the charged surface becomes DEPOLARIZED or neutralized. An impulse permits the __positive__ charges on the outer surface and _____ charges on the inner surface of a neuron to unite.

positive
(+)

negative
(−)

7-306. Impulses increase the __permeability__ of the membrane surrounding a nerve fiber resulting in a __depolarized__ or neutralized fiber.

permea- bility depolarized	**7-307** The depolarization action "runs" along the neuron until the impulse reaches the nerve endings. You should remember that the transmission of an impulse along a nerve fiber is an ___electrical___ phenomenon; the transmission of an impulse from one neuron ending to the next neuron, or organ, is a chemical phenomenon.
electrical	**7-308** At the nerve endings of: –All parasympathetic and sympathetic preganglionic fibers, –All parasympathetic postganglionic fibers, sympathetic postganglionic fibers to sweat glands, –Motor fibers to skeletal muscles, a chemical known as ACETYLCHOLINE (a-SE-til-KOL-een) is produced. (No response required, move to frame 309).
	7-309 A chemical such as (1) _acetylcholine_ ,which is released at (2) ___all___ (all/most) parasympathetic post-ganglionic endings, induces a (3) _chemical_ reaction permitting an impulse to pass from one neuron to the next.
1. acetylcholine 2. all 3. chemical	**7-310** Acetylcholine, which induces impulse transmission between neurons, is produced by the nerve endings of: –all (1) _preganglionic_ sympathetic and parasympathetic fibers –all parasympathetic (2) _postganglionic_ fibers –sympathetic postganglionic fibers going to (3) _sweat glands_ –motor fibers to (4) _skeletal_ muscles.
1. pre- ganglionic 2. post- ganglionic 3. sweat glands 4. skeletal	**7-311** Impulse transmission between neurons is a _chemical_ phenomenon. Impulse transmission along a neuron is an _electrical_ phenomenon.

chemical electrical	**7-312** An all-inclusive statement about nerve impulse transmission would be one indicating that impulse transmission is a combined _electro_ - _chemical_ phenomenon.
electro- chemical	**7-313** The acetylcholine produced at the nerve endings is rapidly destroyed (in one thousandth of a second) by an enzyme called <u>CHOLINESTERASE</u> (KOL-in-ES-ter-ase). The enzyme - _cholinesterase_ - destroys acetylcholine.
cholinesterase	**7-314** After a nerve impulse is transmitted from one neuron to the next, the _acetylcholine_ that induced the transmission is destroyed by the enzyme _cholinesterase_
acetylcholine cholinesterase	**7-315** Nerve fibers that <u>produce</u> acetylcholine are <u>called CHOL-INERGIC</u> (KOL-en-ER-gic). Post-ganglionic sympathetic sweat gland nerves would be classified as _cholinergic_ nerve fibers.
cholinergic	**7-316** All parasympathetic and sympathetic pre-ganglionic fibers are also _cholinergic_
cholinergic	**7-317** With the exception of the postganglionic sympathetic sweat gland fibers, all sympathetic postganglionic fibers produce a chemical known as <u>EPINEPHRINE</u> (EP-i-NEF-rin). Efferent sympathetic fibers to the heart would produce _epinephrine_ at their nerve endings.
epinephrine	**7-318** Epinephrine = ADRENALIN (ad-REN-al-in) Norepinephrine = Noradrenalin Epinephrine and norepinephrine (and their related compounds) are many times referred to as _adrenalin_ and _noradrenalin_ respectively.

259

adrenalin nor- adrenalin	**7-319** All ___*post*___ (pre-ganglionic/postganglionic) sympathetic nerve endings, excepting those afferent sympathetic fibers that innervate sweat glands, produce ___*epinephrin*___ at their nerve endings.
post- ganglionic epinephrine	**7-320** Nerve fibers that produce epinephrine are called ADRENERGIC (ad-REN-er-gic). Most postganglionic sympathetic fibers are ___*adrenergic*___ (cholinergic/adrenergic).
adrenergic	**7-321** All preganglionic fibers of both the parasympathetic and sympathetic divisions produce ___*acetylcholine*___ at their nerve endings and are therefore referred to as ___*cholinergic*___ nerve fibers.
acetyl- choline cholinergic	**7-322** Acetylcholine is destroyed by the enzyme, ___*cholinesterase*___ which is present at the cholinergic nerve endings.
cholinesterase	**7-323** Most (1) ___*post*___ sympathetic nerve fibers produce the chemical (2) ___*epinephrin*___. These nerves are called (3) ___*adrenergic*___ fibers.
1. post- ganglionic 2. epinephrine 3. adrenergic	**7-324** A nerve in a resting state has ___*+*___ ions on the outside of the membrane that surrounds the nerve fiber. This fiber is polarized if there are an equal number of ___*− ions*___ on the inner surface of the neuron membrane.
positive negative ions	**7-325** The ___*permeability*___ of a nerve fiber increases when stimulated, and the nerve fiber becomes ___*depolarized*___.

permeability depolarized	7-326 The electrical depolarization action "runs" along the neuron until the impulse reaches the nerve ___*ending*___, where the impulse is transmitted to the next neuron or organ by a ___*chemical*___ reaction.
endings chemical	7-327 An impulse from a motor fiber is transmitted to a skeletal muscle by way of ___*an*___, which is produced at the nerve endings.
acetylcho- line	7-328 A stimulated nerve returns to a resting state when the ___*agent*___ produced at the nerve endings is destroyed by the ___*enzyme*___, cholinesterase.
acetylcho- line enzyme	7-329 Epinephrine and norepinephrine are also produced by the MEDULLA (me DUL-a) of the ADRENAL (ad-RE-nal) glands of the body. There are two adrenal glands – one above and near each kidney. The effect of epinephrine on an organ, whether it is produced by ___*adrenergic*___ nerve fibers or the ___*adrenal*___ glands, is the same.
adrenergic adrenal	7-330 Postganglionic sympathetic fibers of the ANS tend to increase the body ___*defense*___ against external or internal changes by producing ___*epinephrine*___
defenses epinephrine	7-331 The SYMPATHOADRENAL (SIM-pa-tho-ad-RE-nal) AXIS refers to the combined effect of the ___*sym*___ nerves and ___*adrenal*___ glands to produce epinephrine for body defenses.
sympathetic adrenal	7-332 Epinephrine is produced by the sympathetic nerve endings and the adrenal glands, which are known collectively as the ___*sympatho*___-___*adrenal*___ axis.

sympatho- adrenal	7-333 The production of epinephrine by these two major sources is achieved if the _sympath_ _adrenal_ is stimulated.

---

sympatho- adrenal axis	7-334

Another example of the functional integration of the nervous system with other body systems is the Hypothalamic-PITUITARY (pi-TU-i-ter-i) Axis. You will recall that the hypothalamus is a group of nuclei located beneath the _thalamus_.

---

thalamus	7-335 The hypothalamus is involved with the relaying of sight and smell impulses, regulation of body temperature, and in concert with the limbic system, correlates and regulates the peripheral motor activity of the _ANS_ (CNS/ANS).

262

7-336

CORPUS CALLOSUM

THALAMUS

CERBELLUM

PITUITARY GLAND

HYPOTHALAMUS

PONS

MEDULLA OBLONGATA

The hypothalamic nuclei also influence the _Pit_____ gland, which is located below the hypothalamic region of the brain.

---

pituitary

7-337   Stimulation of the hypothalamic-_____axis would produce the combined effects of both the hypothalamus and the pituitary gland.

---

pituitary

7-338   The sympathoadrenal axis and the hypothalamic-pituitary axis are illustrations of the _____ _____ of the nervous system with other body systems.

---

functional integration

7-339   The pituitary gland and the adrenal glands are parts of the ENDOCRINE (EN-do-krin) System. The endocrine system will be the next chapter, Chapter 8, of this course which you will study.

Part 1

The Endocrine System

8-1   In past units, you have read about systems composed of various organs of the human body; the system which is the subject of this unit is composed of glands. The glands of the body are divided into two categories: the EXOCRINE (EX-o-krin) glands and the ENDOCRINE (END-o-krin)_____ .

---

glands

8-2   Glands, you remember, are body organs which produce and secrete important substances. Glands are classified as "endocrine" or "exocrine" according to where they _____ the particular substances they produce.

---

secrete

8-3   The characteristics of exocrine glands are that they have tubes called ducts which carry their secretions to the outside of the body or into hollow organs. Endocrine glands, on the other hand, do not have ducts, but secrete their substances into the blood stream. To express this distinction, the _____ glands are also referred to as the "ductless" glands.

---

endocrine

8-4   The glands which you have encountered as part of the integumentary system and elsewhere -- sweat, mammary, sebaceous glands etc. -- all secrete through ducts and are therefore exocrine glands.

   This unit covers the other category, the ductless _____ glands which compose the _____ system.

---

endocrine
endocrine

8-5   The secretions of endocrine glands are called HORMONES (HOR-moans). Hormones are first formed and secreted by the endocrine glands and then transported to various parts of the body by means of (ducts/the bloodstream) _____ .

the blood- stream	8-6	Hormones can be defined as chemical substances which are produced by gland cells in one part of the body and travel through the bloodstream to affect, in most cases, the function of cells in (another/the same) _____ part of the body.
another	8-7	The part of the body whose cells are affected by the hormones from a given endocrine gland is called the TARGET ORGAN. Through the agency of the hormones it produces, then, an endocrine gland exercises a kind of chemical <u>remote</u> <u>control</u> over its particular _____ organ or organs.
target	8-8	It is this remote control function exercised by the various _____ glands over their respective _____ _____ which accounts for the importance of the <u>endocrine</u> <u>system</u>.

ENDOCRINE GLANDS

PITUITARY GLAND
(Hypophysis cerebri)

THYROID GLAND

PARATHYROID GLANDS

ADRENAL GLANDS

Pancreas, containing ISLANDS OF LANGERHANS

(Kidney)

OVARIES (FEMALE)

TESTES (MALE)

endocrine target organs	8-9	Look at the drawing of the endocrine system which shows the locations of the glands in the endocrine system. As you can see, these glands are (compactly arranged/ widely separated) _____ in the body.
widely separated	8-10	The glands in the endocrine system are not attached to each other in any way; they are related not in structure but in function. The endocrine system consists of a group of glands all of which exercise the function of chemical remote _____.
control	8-11	The endocrine system, along with the nervous system, is one of the two major regulatory or _____ systems in the human body.
control	8-12	In general, the hormones of the endocrine glands act to regulate the various metabolic processes of the body. It should be understood that "metabolism" simply means "change;" in its broadest sense, the term metabolism is used to include all the chemical and physical _____ involved in the development and functioning of the living body.
changes	8-13	Thus, for example, the electrolytic changes undergone in the body by the essential minerals such as calcium or sodium, the changes undergone by the nutrients such as the carbohydrates, the gross changes undergone by the tissues themselves -all these are termed _____ processes.
metabolic	8-14	The maintenance of normal growth, development, and function of the adult human body is therefore dependent on the presence and proper functioning of all the endocrine _____ in the endocrine _____.
glands system	8-15	Now let's examine the individual endocrine glands in detail. Look at the drawing again; the large single gland located in the neck is called the _____ gland.

thyroid	8-16	The THYROID (THI-roid) gland is located in the front part of the neck; it consists of two lateral portions or lobes joined in front by a narrow band of tissue.

The arrows on the diagram point to the _____ of the thyroid gland.

THYROID CARTILAGE

THYROID GLAND

TRACHEA

FRONT VIEW

lobes	8-17	The function of the thyroid gland is to secrete a hormone which regulates the rate of metabolism of human body cells. The building up and, to a somewhat lesser extent, the breaking down of protoplasm throughout the entire body is influenced by the hormone secreted by the _____ gland.

thyroid	8-18	A normal rate of metabolism is an important part of normal body function, most particularly growth. The thyroid hormone, since it regulates the _____ of metabolism of cells throughout the body, is thus essential to maintaining _____ body function.

rate normal	8-19	The body, however, can continue to function without the thyroid hormone even if it does not function normally. In other words, although the thyroid hormone is essential to normal daily activity by virtue of its influence on the rate of _____, it is not essential to _____.

metabolism life	8-20	The hormone secreted by the thyroid gland is called THYROXIN, each molecule of which contains four atoms of iodine. The atoms of iodine are the most important components of _____.

thyroxin	8-21	Iodine is ingested with the food we eat, is circulated in the bloodstream, and is absorbed by epithelial cells in the thyroid gland, where it is then used to manufacture the thyroid hormone _____.

thyroxin	8-22	The importance of iodine to thyroxin is reflected by several special characteristics of the thyroid gland. The epithelial cells in the thyroid, for instance, are more efficient at extracting iodine from the system than other tissue in the body. The result of this is that practically all of the _____ in the body is concentrated in the _____ gland.
iodine thyroid	8-23	Again, if the system is deficient in iodine for any reason, the thyroid gland will enlarge itself in an effort to trap the available iodine more effectively. When the thyroid is thus _____ ,it is called a goiter.
enlarged	8-24	When the thyroid gland is functioning as it should be, it is producing just the amount of thyroxin necessary to maintain a normal rate of metabolism for the rest of the body. Since the body's metabolic rate can be measured, it is relatively easy to determine whether or not the thyroid gland is _____ properly.
functioning	8-25	The rate of metabolism is measured as follows: The chemical activity involved in cell metabolism produces heat. An individual's metabolic rate can be determined by measuring the amount of _____ being produced by metabolic activity within his body.
heat	8-26	However, since other factors besides metabolism also produce heat -- mental and physical activity, the ingestion of food-- the individual is asked to avoid these activities as far as possible, so that the heat in his body is being produced almost solely by _____ activity.
metabolic	8-27	An individual who is refraining from mental and physical activity and the ingestion of food is described as being in a resting or BASAL (BAY-sl) condition, and the measurement of his heat production in that condition is called his _____ Metabolic Rate or B.M.R.
basal	8-28	An individual who has avoided (1) _____ and (2) _____ activity and the ingestion of (3) _____ is described technically as being in a (4) _____ condition.

1. mental 2. physical (either order) 3. food 4. basal	8-29	The B.M.R. or _____ _____ _____ can give a good indication of whether or not the _____ gland is functioning properly.
Basal Metabolic Rate thyroid	8-30	When the thyroid gland is secreting an abnormal amount of thyroxin, the B.M.R. rises or falls accordingly. An abnormally high B.M.R., for instance, would result from an excessive amount of metabolic activity in the body and would indicate that the thyroid was producing (too much/too little) _____ thyroxin.
too much	8-31	An excess of thyroid hormone in the body is called HYPERTHYROIDISM. "Hyper" means "too much," as in "Hyperactive" or "hypersensitive." An abnormally high metabolic rate is a reliable sign of _____ thyroidism.
hyper	8-32	When metabolic activity is intensified by the presence of too much thyroxin, the body may be compared to an automobile engine with the idling speed set too high; the body, like the engine, produces more waste, burns more fuel, and produces more heat than is normal for it. All of these are characteristics of the bodily condition called _____.
hyper- thyroidism	8-33	The thyroid gland may also produce too little thyroid hormone, with a resultant slowing down of the metabolic rate. The term for this condition is HYPOthyroidism. Just as the prefix "hyper" means "too much," the prefix "_____" means "too little."
hypo	8-34	When the metabolic rate slows down as a result of insufficient thyroid hormone (caused by a diseased thyroid or a lack of iodine in the system), the body has a tendency to produce less heat and consume less fuel; these signs are characteristics of the condition called _____.
hypo- thyroidism	8-35	An individual suffering from hypothyroidism is likely to feel (cold/hot) (1) _____; also, since his appetite may be unaffected but he is burning (more/less) (2) _____ fuel, he would tend to (gain/lose) (3) _____ weight.

1. cold 2. less 3. gain	8-36	An individual with hyperthyroidism, on the other hand, would have a feeling of _____ in his body and would tend to _____ weight.
hotness, heat lose	8-37	Hyper- and hypothyroidism are not usually fatal. If they are not treated, however, they can be damaging to _____ growth and body function.
normal	8-38	For this reason the thyroid hormone thyroxin is particularly important in the periods of rapid _____ during infancy and youth.
growth	8-39	Let's review briefly what you've learned so far. The thyroid gland is a part of the (1) _____ system. This system, with the nervous system, is one of the two major (2) _____ systems of the body. In general, the glands of this system act to regulate the (3) _____ processes of the body.
1. endocrine 2. control 3. metabolic	8-40	All endocrine glands secrete into the _____. The secretions of all endocrine glands are called _____.
bloodstream hormones	8-41	The thyroid hormone is called _____. The most important components of the thyroid hormone are its atoms of _____.
thyroxin iodine	8-42	Thyroxin regulates the (1) _____ of _____ of the cells throughout the body. For his thyroid activity to be measured, an individual should be in a (2) _____ condition. The measurement made when he is in that condition is called the (3) _____ _____ _____ (initials).

1. rate (of) metabolism 2. basal 3. B.M.R.	8-43	An abnormally high B.M.R. indicates a condition of (1) _____, or (too much/too little) (2) _____ thyroid hormone in the body. An abnormally low B.M.R. indicates a condition of (3) _____.
1. hyperthy-roidism 2. too much 3. hypothy-roidism	8-44	Thyroid hormone is essential to _____ (your own words), but not essential to _____.
normal growth and body function life	8-45	Now refer to the Endocrine System drawing again. The small glands which appear to be imbedded in the larger thyroid gland (but may be either embedded in the thyroid or simply very near to it, depending on the individual) are the _____ glands.
parathyroid	8-46	The parathyroid glands are each about the size of a pea; they are usually four in number, two associated with each lobe of the thyroid. Arrow A points to two _____ glands. Arrow B points to one _____ of the thyroid.
parathyroid lobe	8-47	The product of the parathyroid is called PARATHORMONE. You should have no difficulty remembering that the (1) _____ – thyroid glands produce an endocrine secretion or (2) _____ called (3) _____.
1. para 2. hormone 3. para-thormone	8-48	Unlike the thyroid hormone thyroxin, the parathyroid hor-mone, _____, is not only essential to normal body function, but also essential to life.

In frame 8-46 diagram: THYROID CARTILAGE — A — B — TRACHEA — BACK VIEW

para- thormone	8-49 The most important metabolic effect of parathormone is on the essential mineral <u>calcium</u>; the electrolytic changes undergone by _____ in the body are profoundly influenced by the parathyroid hormone.
calcium	8-50 The primary function of the parathyroid glands is to maintain a normal concentration of calcium in the blood and tissue fluids. Despite wide variations in calcium ingestion, the parathyroids control the levels of (1) _____ in the (2) _____ and tissue (3) _____ within very narrow limits.
1. calcium 2. blood 3. fluids	8-51 In order to perform this function, parathormone acts principally in two areas. Most of the calcium in the body is stored in the bones; parathormone mobilizes calcium out of the (1) _____ and into the (2) _____ and (3) _____ fluids.
1. bones 2. blood 3. tissue	8-52 At the same time, parathormone acts to move calcium out of the blood again by promoting its excretion through the kidneys. Thus, the two principle sites of parathormone's action are the bones from which calcium is mobilized and the _____ from which calcium is excreted in the urine.
kidneys	8-53 The effects of parathormone are therefore very widespread in the body, as may be seen particularly when the supply is abnormal. An oversupply of parathormone results in: Increased decalcification of the (1) _____; Increased excretion of (2) _____ by the (3) _____.
1. bones 2. calcium 3. kidneys	8-54 The most important result of an oversupply of parathormone, however, is the (increased/decreased) _____ concentration of calcium in the blood and in the fluid of _____ throughout the body.
increased tissues	8-55 Calcium plays a role not only in the mineralization of bones but also in many other body functions, including blood coagulation, the irritability of skeletal muscles, and the rhythm of the heart; and changes in blood and tissue fluid calcium concentration may affect all these functions. (No answer required)

	8-56	Thus an overproduction of parathormone may not only demineralize and weaken the bones, but also can lead ultimately to kidney and circulatory collapse; this condition would be called (hyper/hypo) _____ parathyroidism and can be fatal if untreated.
hyper	8-57	The opposite situation -- an undersupply of parathormone and resultant drop in the level of calcium in the blood and tissue fluids -- is called _____ and can also be fatal.
hypopara- thyroidism	8-58	Hypoparathyroidism can produce increased muscular and nervous irritability, leading to twitching, cramps, convulsions, and ultimate paralysis of the respiratory system if untreated. These symptoms are the result of an (oversupply/undersupply) _____ of parathormone.
undersupply	8-59	Furthermore, in addition to its effect on the metabolism of calcium, parathormone also influences the metabolism of phosphorus. Most of the phosphorus in the body (about 80%) is found in combination with calcium in the bones and teeth, and the metabolism of _____ is closely connected with that of calcium.
phosphorus	8-60	Thus although the principle metabolic effect of parathormone is on calcium, it also affects the metabolism of _____.
phosphorus	8-61	To review: The vital function of the parathyroid hormone, (1) _____, is to regulate the level of (2) _____ in the (3) _____ and (4) _____ fluids.
1. para- thormone 2. calcium 3. blood 4. tissue	8-62	It performs this function primarily by mobilizing calcium out of the _____ and into the blood and tissue fluids, and at the same time by promoting the removal of calcium from the blood for excretion in urine by the _____.

bones kidneys	8-63 In addition to its effect on the metabolism of calcium, the hormone secreted by the _____ glands also influences the metabolism of _____.

parathyroid
phosphorus
	8-64 <u>ADRENAL GLANDS</u>
	     The next endocrine glands to be studied are the ADRENAL (ad-REEN-al) glands.
	     Look at the Endocrine System diagram. There are _____ (number) adrenal glands in the body.

two	8-65 The prefix "ad" means "near" and the word "renal" means "kidney." The location of the two _____ glands is "near the kidneys," one above each kidney.
adrenal	8-66 Each adrenal gland has two distinct parts, the MEDULLA (meh-DULL-ah) and the CORTEX (KOR-tex). The cortex is a pale pink envelope of tissue encasing the dark brown inner core, the _____.
medulla	8-67 Both the medulla and the cortex of an adrenal gland secrete hormones; however, the hormones secreted by the inner core, the _____, are quite different from those secreted by the outer envelope, the _____.
medulla cortex	8-68 The adrenal <u>medulla</u> secretes two hormones which you have already encountered in the preceding unit. The first is called EPINEPHRINE or ADRENALINE; "epinephrine" is the preferred term. You should, however, remember that epinephrine and _____ are the same hormone.
adrenaline	8-69 The second adrenal medullary hormone is called NOREPINE-PHRINE (or noradrenaline). Although norepinephrine differs somewhat from epinephrine both in nature and in degree of activity, their effects are so closely related that epinephrine and _____ are commonly referred to together as "the epinephrines."

274

norepine- phrine	8-70	For the sake of simplicity, the following frames will use the term <u>epinephrines</u> when describing the activities of both hormones -- epinephrine and _____ -- secreted by the adrenal _____.
norepine- phrine medulla	8-71	In the preceding unit you learned that the sympathetic nervous system releases an epinephrine-like substance used as a transmitting agent. The effects of the _____ nervous system on the body--increase of heart rate, blood pressure, mental alertness, and blood sugar content--can also be described as the effects of the adrenal medullary hormones, the _____.
sympathetic epinephrines	8-72	The epinephrines are often called the "emergency" hormones because they mobilize what is termed the "fight or flight mechanism." The production of the epinephrines is stimulated by such strong emotions as fear, rage, etc. for the purpose of helping the body meet-- by _____ or fleeing -- whatever _____ caused these emotions.
fighting emergency	8-73	The epinephrines put the whole body in a state of readiness for an emergency by increasing mental alertness, raising the level of sugar in the blood (as fuel for muscles) and increasing _____ pressure and the rate at which the _____ beats, for maximum efficiency.
blood heart	8-74	Thus the epinephrines play an important role in helping the body adjust to its environment. The body can, however, survive without the epinephrines (in a protected environment) --in other words, they are not _____ to life.
essential	8-75	To review: You should remember that the <u>two</u> hormones secreted by the adrenal (1) _____ -- the inner core of the adrenal gland-- are (2) _____ and (3) _____ also called adrenaline and noradrenaline.
1. medulla 2. epinephrine 3. norepine- phrine	8-76	Epinephrine and norepinephrine are secreted by the medulla in response to strong emotions caused by an _____, and they act to mobilize the _____ or _____ mechanism.

emergency fight (or) flight	8-77	The epinephrines prepare the body to meet an emergency by increasing: (1) mental _____; (2) the _____ content of the blood; (3) blood _____; and (4) the rate at which the _____ _____ .
1. alertness 2. sugar 3. pressure 4. heart beats	8-78	The adrenal medullary hormones (are/are not) _____ essential to life.
are not	8-79	On the other hand, the hormones secreted by the <u>outer en-velope</u>, the _____, of the adrenal glands <u>are</u> essential to life.
cortex	8-80	The adrenal cortex secretes a number of hormones called, collectively, the ADRENAL CORTICO-STEROIDS. The part of the name which refers to their origin in the <u>cortex</u> is "_____." This part is often omitted for brevity, making them simply the "_____ _____."
cortico adrenal ster- oids	8-81	"Steroids" are chemical compounds of the lipid class and are related to each other in that they all have the same chemical nucleus. As a result, the different adrenal _____ (or adrenal _____ - _____) have many functions in common.
steroids cortico- steroids	8-82	The most important functions common to the majority of adrenal cortico-steroids are the regulation of salt metabolism and of carbohydrate (sugar) metabolism. The adrenal steroids vary first of all according to whether their influence is exerted chiefly over the metabolism of _____ or the metabolism of _____ .
salt carbohydrate (either order)	8-83	Those adrenal steroids whose primary effect is on carbo-hydrate metabolism are termed GLUCOCORTICOIDS. Cells in various parts of the body -- notably liver, muscle, and fat cells-- are acted on most particularly by the _____ to convert fat and protein into _____ .

gluco- corticoids carbohydrate (or sugar)	8-84	The single adrenal steroid hormone which shows the greatest effect on carbohydrate metabolism is HYDROCORTISONE. As the glucocorticoids are the most potent carbohydrate metabolizers among the adrenal steroids, so _____ is the most potent glucocorticoid.
hydrocor- tisone	8-85	The levels of blood sugar in the body, the depositions of fat, the life span of white blood cells, and a number of other vital body activities are affected to an important degree by the _____ of the adrenal cortex and most particularly by the hormone _____.
gluco- corticoids hydro- cortisone	8-86	Another adrenal steroid of the glucocorticoid category which should be singled out for attention is CORTISONE. You have probably heard of the treatment of arthritis and also various skin disorders similarly typified by inflammation by the use of _____.
cortisone	8-87	Although it exerts an important influence on carbohydrate metabolism like all the glucocorticoids, it is particularly for its anti-inflammatory powers that _____ is noted.
cortisone	8-88	The second category of adrenal steroid hormones includes those which exert their effect primarily on salt metabolism; these hormones are termed MINERALOCORTICOIDS. The absorption and retention of _____ by cells throughout the body are profoundly influenced by the _____ of the adrenal cortex.
salt mineralo- corticoids	8-89	One hormone in this category, called ALDOSTERONE, is estimated to be at least twenty-five times as effective as any other adrenal steroid in the retention of salt. Thus as hydrocortisone is the most potent glucocorticoid, so _____ is the most potent _____.
aldosterone mineralo- corticoid	8-90	As a result of their influence on _____ metabolism, the mineralocorticoids also exert far-reaching effects on electrolyte distribution, water balance, and blood pressure; and in all of these areas, _____ is the adrenal steroid hormone with the greatest potency.

salt aldosterone	8-91	Again, the mineralocorticoids and particularly aldosterone also promote the excretion of potassium from the kidneys. If, as a result of inadequate mineralocorticoid activity, the kidneys fail to excrete the normal quantity of _____, the effects produced on the nervous and cardiovascular systems can be fatal.
potassium	8-92	It should be pointed out that <u>all</u> the adrenal cortico-steroids act to promote salt retention and _____ excretion to a greater or lesser degree.
potassium	8-93	The third and final category of adrenal steroids is a group of <u>sex</u> <u>hormones</u> that are similar in structure and action to the hormones secreted by the sex glands, which will be covered later in the unit. The _____ hormones of the adrenal cortex differ from those of the sex glands in being substantially less potent.
sex	8-94	It is worth noting, however, that degree of effectiveness is the <u>only</u> major difference between the _____ _____ secreted by the adrenal cortex and those secreted by the sex glands.
sex hor- mones	8-95	Now a brief review. The hormones secreted by the adrenal cortex--although they share most of their functions to a greater or lesser degree-- are divided into 3 categories according to primary effect. These categories are: (1) the _____; (2) the _____; (3) the _____ _____;

1. glucocorti-
    coids
2. mineralo-
    corticoids
(either order)
3. sex hormones
	8-96	The glucocorticoids chiefly influence (1) _____ metabolism. Two glucocorticoids of particular importance are: (2) _____, because it is the most potent; and (3) _____ , because of its anti-(4) _____ powers.

1. carbohy- drate (or sugar) 2. hydrocorti- sone 3. cortisone 4. inflammatory	**8-97** The mineralocorticoids chiefly influence (1) _____ metabolism; the most potent member of this group is (2) _____. The sex hormones of the adrenal cortex differ from those of the sex glands chiefly in (3) _____ of effect- iveness.
1. salt 2. aldosterone 3. degree	**8-98** The hormones of the adrenal cortex are called, collectively, the (1) _____ - _____, or more simply the (2) _____ _____. The hormone epinephrine, also called adrenaline, is se- creted by the adrenal (3) _____.
1. adrenal cortico- steroids 2. adrenal steroids 3. medulla	**8-99** PANCREAS Refer to the Endocrine System diagram. The PANCREAS (PAN-kree-as) is a gland located in the abdomen which secretes digestive juices into the small intestine through a duct. An in- tegral part of the pancreas is a group of specialized cells which secrete a hormone directly into the bloodstream. The pancreas as a whole, then performs (endocrine/exocrine/both endocrine and exocrine) _____ functions.
both endo- crine and exocrine	**8-100** The specialized cells within the pancreas which secrete directly into the bloodstream are called the ISLANDS OF LANG- ERHANS (LANG-er-hans). The <u>endocrine</u> functions of the pancreas are performed by the _____ _____ _____.
islands of Langerhans	**8-101** The drawing shows a magnified section of the pancreas with an _____ _____ _____, drawn in red, embedded in it.

island of Langerhans	8-102	The islands of Langerhans produce a hormone essential to life known as <u>INSULIN</u>. Insulin is indispensable to normal carbohydrate metabolism. Carbohydrate metabolism is regulated by the hormones of the adrenal cortex and medulla, by the thyroid, and by a hormone called _____.
insulin	8-103	Insulin is the secretion of the _____ _____ _____ located in the _____.
islands of Langerhans pancreas	8-104	Insulin is a complex protein which works at the cellular level to enable cells to take up carbohydrates in order to convert them for energy. Without insulin, _____ metabolism all but ceases at the _____ level.
carbohydrate cellular	8-105	When they cannot be metabolized, the now useless carbohydrates accumulate in the bloodstream, raising the amount of sugar in the _____ to a dangerous level.
blood, bloodstream	8-106	The excess sugar spills over through the kidneys into the urine, carrying with it a large amount of water. The loss of the _____ which is carried along by the excess _____ causes the body to become dehydrated.
water sugar	8-107	The disease characterized by dehydration of the body, resulting in excessive thirst, and by excessive production of urine containing sugar, is familiar to you as diabetes mellitus. If untreated by means of insulin injections, _____ _____ can lead to coma, convulsions and death.
diabetes mellitus	8-108	At the same time, if a lack of insulin prevents the cells from producing energy by converting carbohydrates, other substances -- namely <u>fat</u> and <u>protein</u> -- must be broken down to produce _____ for the body.

energy	8-109 The breaking down of fat and protein to produce energy also results in the production of certain <u>acids</u> which enter the bloodstream and raise the acidity of the <u>blood</u>. The production of these _____, as a side-effect of a (lack/excess) _____ of insulin, can be fatal if unchecked.
1. acids 2. lack	8-110 Finally, an <u>excessive</u> supply of insulin causes a (drop/rise) _____ in the level of blood sugar, which can lead also to convulsions and death.
drop	8-111 To review briefly: The endocrine functions of the pancreas are performed by the _____ _____ _____, which secrete the hormone _____.
islands of Langerhans insulin	8-112 Insulin is indispensable to (1) _____ metabolism at the cellular level and therefore essential to maintaining a healthy level of (2) _____ in the blood. Insulin is clearly also essential to (3) _____.
1. carbohy- drate 2. sugar 3. life	8-113 GONADS The GONADS (GO-nads) are the sex glands of the body. <u>Both</u> the male and the female sex glands are called _____.
gonads	8-114 In the male, the sex glands are the TESTES (TEST-eez), two oval glands contained in a sack of skin suspended between the legs. The singular of _____ is TESTIS. These two glands are the sex glands or _____ of the male.
testes gonads	8-115 The male sex glands produce the male sex hormones, the most important of which is called TESTOSTERONE (tes-TOST-er-own), a steroid. The adrenal cortico-steroids and the hormone _____ secreted by the male sex glands, the _____, are members of the same chemical family.

testo- sterone testes	8-116	Embedded in the testes are a number of specialized interstitial cells, and these are the cells directly responsible for the production of the hormone _____.
testo- sterone	8-117	Testosterone stimulates the development of male secondary sex characteristics. The primary sex characteristics of the male -- presence of a penis and testicles -- are not controlled by the testes. Testosterone is therefore essential only to the development of _____ sex characteristics in the male.
secondary	8-118	Testosterone is responsible for such characteristics as the development of the beard, growth and distribution of body hair, and deepening of the voice. All these are classified as _____ sex _____.
secondary character- istics	8-119	Testosterone is also responsible for the "sex drive" or LIBIDO (li-BEE-doe) of the male. A eunuch -- a male deprived of testosterone, for instance, by the removal of the testes by castration -- suffers the loss both of _____ _____ _____ and of _____.
secondary sex characteris- tics libido	8-120	The testes are not only endocrine glands. They also perform an exocrine function, namely, in the production of SPERM cells. SPERM is the name for the reproductive _____ of the male, which are, of course, carried to the outside of the body.
cells	8-121	Sperm is produced by the cells of a mass of minute tubes or tubules coiled within the testes. These tubules are connected to a system of EXCRETORY DUCTS. The _____ cells secreted by the testes are carried to the outside of the body by the system of _____ _____, which terminates in the penis.
sperm excretory ducts	8-122	Thus the testes perform not only an endocrine function, the secretion of the hormone (1) _____, but also an exocrine function, the production of (2) _____ cells, which are the (3) _____ cells of the male.

1. testosterone 2. sperm 3. reproductive	8-123 The gonads of the <u>female</u> are the OVARIES (OH-va-reez). The ovaries are two small bean-shaped masses of tissue located in the lower part of the abdominal cavity. Like the testes, the _____ have both exocrine and endocrine functions.
ovaries	8-124 The ovaries produce two hormones. One is called "progesterone" and will be discussed later, in the section on reproduction. The other is called ESTROGEN (ES-tro-gen). Like testosterone in the male, estrogen is responsible for the (primary/secondary) _____ sex characteristics of the female.
secondary	8-125 Estrogen controls the female libido, and development of the breasts, widening of the hips, distribution of body fat, onset of menstruation, and other secondary sex characteristics of the _____ body. Again, like testosterone, _____ belongs to the chemical family of steroids.
female estrogen	8-126 The exocrine function of the ovaries is the development and expulsion of the egg or OVUM (O-vum), the female counterpart of the sperm. The <u>female</u> _____ cell is the egg or _____.
reproductive ovum	8-127 The ova (plural of ovum) are produced by specialized cells. These epithelial cells group together into islands, each island producing an _____ at its center.
ovum	8-128 These islands are called FOLLICLES (FOL-ik-ls). The whole island, including epithelial cells and the _____ which they produce, is included in the term _____.
ovum follicle	8-129 You should remember that the production of ova within the cell islands or (1) _____ of the ovaries is an (endocrine/exocrine) (2) _____ function comparable to the production of (3) _____ by the testes.

1. follicles
2. exocrine
3. sperm
	8-130 PITUITARY GLAND
	
	       The last member of the endocrine system to be studied is the PITUITARY (pi-TU-i-terry) GLAND whose Latin name is "hypophysis cerebri."
	
	       Look at the Endocrine System diagram. The pituitary is a compact mass of tissue about the size of a pea, located in the _____.

head or skull	8-131 The pituitary gland is contained in a small cavity in the floor of the skull, directly behind the bridge of the nose. The pituitary is a double gland. It consists of two lobes, like the thyroid, but each pituitary lobe is considered a separate _____ gland.
pituitary	8-132 The lobe of the pituitary toward the back side or posterior of the skull is called the posterior pituitary gland. The lobe toward the front, or anterior, of the skull is called the _____ _____ _____.
anterior pi- tuitary gland	8-133 Both the lobe toward the back of the skull, the _____ pituitary gland, and the lobe toward the front of the skull, the _____ _____ gland, secrete several distinct hormones.
posterior anterior pi- tuitary	8-134 The pituitary gland has the title of "master gland of the endocrine system." It owes this title to a group of hormones secreted by the anterior pituitary, whose function is to regulate the productivity of other glands of the _____ system.
endocrine	8-135 These hormones are called the TROPIC hormones, the word "tropic" meaning "affecting development." The healthy development and functioning of several of the endocrine glands which you have studied is under the direct control of the _____ hormones secreted by the _____ pituitary.

tropic anterior	8-136	Thus the several GONADOTROPIC hormones of the anterior pituitary regulate the growth and proper functioning of the male and female _____.
gonads	8-137	Both the endocrine and exocrine functions of the testes and ovaries are stimulated by the several _____ hormones, as you will see in more detail in the second part of this chapter.
gonado- tropic	8-138	Another tropic hormone secreted by the anterior pituitary is called ADRENOCORTICO-TROPIC HORMONE, or more simply A.C.T.H. The secretion of hormones by the _____ _____ is stimulated by ___. ___. ___. ___. (initials).
adrenal cortex A.C.T.H.	8-139	You should remember that not only the productivity but also the healthy growth and development of the adrenal cortex is dependent on ___. ___. ___. ___.
A.C.T.H.	8-140	The final member of the tropic hormone group is the THYROTROPIC hormone, which stimulates the growth of, and production of thyroxin by, the _____ gland.
thyroid	8-141	Thus the tropic hormones of the anterior pituitary include: the several (1) _____ hormones; (2) ____ ____ ____ ____ ; and the (3) _____ hormone.
1. gonado- tropic 2. A.C.T.H. 3. thyrotropic	8-142	The way in which the pituitary's own activity is regulated is called a feed-back mechanism . A drop in the production of thyroxin, for instance, alerts the anterior pituitary to produce the hormone which stimulates the thyroid. This is the process described by the term "_____ mechanism."

feedback	**8-143** The pituitary will be stimulated to increase production of thyroid – stimulating hormones by a drop in the level of thyroxin in the body; it will also respond to a rise in the level of thyroxin by (decreasing/increasing) _____ production of thyroid-stimulating hormones. Both of these are examples of the _____ mechanism.
decreasing feedback	**8-144** The anterior pituitary hormones discussed so far are those which control the activities of other glands and thus account for the pituitary's title of "master" gland. The anterior pituitary also produces an important hormone which is called the GROWTH HORMONE because of its direct influence on _____.
growth	**8-145** The proper name for the growth hormone is SOMATOTROPIC hormone. The word "somato" comes from the Greek word "soma", meaning "body"; the combination of "_____" and "tropic" tells you that this hormone affects the growth or development of the _____.
somato body	**8-146** Although it is not the only growth-influencing hormone, its importance to normal body development can be seen in the fact that the rate of growth of an individual may be reduced by as much as 50% in the total absence of _____ (proper name) hormone.
somato- tropic	**8-147** The somatotropic hormone stimulates body growth in general, especially the growth of long bones, and it exerts a particular influence on the metabolism of carbohydrates and proteins. Unlike the other tropic hormones of the anterior pituitary, somatotropic hormone produces these effects (directly/ through its regulation of other endocrine glands) _____.
directly	**8-148** A lack of somatotropic hormone during the body's period of growth will result in a midget: person of normal proportions but unusually small size. Conversely, an excess of somatotropic hormone during the growth period will produce an individual of normal proportions but unusually _____ size.
large	**8-149** A circus giant is a good example of a person whose anterior pituitary gland produced (too much/too little) _____ somatotropic hormone during his growth period.

too much	8-150 Giants are persons of abnormal size but normal proportions. If, on the other hand, an excessive production of somatotropic hormone occurs <u>after</u> the body's normal period of growth, the result will be a person of abnormal _____; outsized hands, feet, and facial bones (diseases referred to as acromegaly) are the most common examples of this condition.
proportions	8-151 Thus, the <u>anterior</u> pituitary secretes both the gland-stimulating tropic hormones, which give the pituitary as a whole its title of " _____ gland of the endocrine system;" it also secretes a hormone essential to proportionate growth and to the rate of growth, the _____ hormone.
master somato- tropic	8-152 Important hormones are also produced by the <u>other</u> lobe of the pituitary gland, the _____ pituitary.
posterior	8-153 The actions of one posterior pituitary hormone, called OXYTOCIN, are restricted to the female body; it stimulates the secretion of _____ from the mammary glands during lactation and stimulates contraction of the female uterus. Obstetricians commonly make use of _____ to induce labor.
milk oxytocin	8-154 By stimulating the (1) _____ of the uterus during childbirth and the (2) _____ of milk from the mammary glands afterwards, the posterior pituitary hormone (3) _____ plays an important role in the total process of reproduction.
1. contraction 2. secretion 3. oxytocin	8-155 The second posterior pituitary hormone is called ANTIDIURETIC HORMONE, or more simply, A.D.H., and exerts its major effect on kidney function. The influence of ___ ___ ___ (initials) will be covered in the second part of this chapter, which examines _____ function in detail.
A.D.H. kidney	8-156 For the time being, you need remember only that the second hormone secreted by the (1) _____ pituitary is called (2) ___ ___ ___ and that this hormone has its major effect on the excretion of urine by the (3) _____.

1. posterior
2. A.D.H. (or
   anti-diuretic
   hormone)
3. kidney
	8-157 The following frames will review briefly the principle hormones of the pituitary gland--whose Latin name is "hypophysis cerebri" and whose title is "_____ gland of the _____ system."

master
endocrine
	8-158 The gland-stimulating hormones of the (1) _____ pituitary are:
	
	     The several (2) _____ hormones;
	     (3) ___. ___. ___. ___. (initials); and the (4) _____ hormone.
	
	     The growth-stimulating (5)_____ hormone is also secreted by this lobe of the pituitary.

1. anterior
2. gonado-
   tropic
3. A.C.T.H.
4. thyro-
   tropic
5. somato-
   tropic
	8-159 The two hormones secreted by the (1) _____ pituitary are (2) _____ and (3) ___. ___. ___. (initials).

1. posterior 2. oxytocin 3. A.D.H.	8-160 Part 11 of this chapter will pay particular attention to the Genito-urinary System and its relationships to the Endocrine System.

8

Part 2 -

The Genito-Urinary System

8-161   The GENITO-URINARY SYSTEM actually consists of two distinct groups of organs, the genital or reproductive organs and the urinary organs. These organs form two distinct systems of their own within the _____ - _____ system.

---

genito-urinary

8-162   It is important to remember that in spite of their different functions, all the organs of both groups are included under the collective title of the _____ - _____ system.

---

genito-
urinary

8-163   THE REPRODUCTIVE SYSTEM

The genital or reproductive organs, which will be considered first, are grouped together to form the REPRODUCTIVE SYSTEM. As its name indicates, the function of this system is to carry out the process of human _____.

---

reproduction

8-164   For the purpose of this course, the human reproductive process will be divided into four successive stages, with which you should be familiar in order to understand the functions of the individual organs of the _____ system.

---

reproductive

8-165   Human reproduction begins with the manufacture of male and female reproductive cells by the male and female sex glands or _____.

---

gonads	8-166	Thus the first stage of the human reproductive process would be the _____ of reproductive cells.
manufacture	8-167	Once the reproductive cells have been produced by their respective gonads, a male and a female cell must be brought together to form the first cell of a new human being. This stage is called FERTILIZATION; a wholly new cell is formed when a female cell is _____ by a male cell.
fertilized	8-168	The second stage of human reproduction is the _____ of a female reproductive cell by a male cell.
fertilization	8-169	Another term for fertilization is CONCEPTION. When the male and female cells unite to form a new cell, this may be described as either fertilization or _____.
conception	8-170	The terms (1) "_____" and (2) "_____" are used interchangeably for the (3) _____ stage of human reproduction.
1. fertilization 2. conception (either order) 3. second	8-171	The new cell formed at conception or fertilization develops within the female body during PREGNANCY. The whole period from conception to birth is termed _____.
pregnancy	8-172	The new member of the species develops from a single cell into a fully formed infant during _____, the _____ stage of the reproductive process.

pregnancy third	8-173 The fourth stage of reproduction is _____ itself, when the child is delivered from the mother.
birth	8-174 Thus the four stages of the reproductive process (as defined for the purpose of this course) are: (1) _____ (your own words), (2) _____ or _____, (3) _____ and (4) _____.
1. manufacture of reproductive cells (or equivalent statement) 2. fertilization (or) conception 3. pregnancy 4. birth	8-175 It should be pointed out that lactation, the manufacture of milk for the new infant by the mother, is also standardly included in the reproductive process, but since the mammary glands, which are responsible for lactation, are not structurally a part of the reproductive system, they will not be studied along with the _____ of that system.
organs	8-176 The organs of the <u>male</u> and <u>female</u> reproductive systems and their respective functions will be examined successively in the following frames. (No answer required)
	8-177 THE MALE REPRODUCTIVE SYSTEM The primary function of the male system is the production of male reproductive cells or sperm. Thus, the organs which produce sperm, the male gonads or _____, are referred to as the _____ organs of the male reproductive system.
testes primary	8-178 The secondary or accessory function of the male system is to provide the means by which the sperm cells can travel to and fertilize the female cell or ovum. The organs which, in their various ways, help the sperm cells _____ to the ovum are referred to as the secondary or _____ reproductive organs.
travel accessory	8-179 The sperm cells, you remember, travel from the testes to the outside of the body through a system of _____.

ducts	**8-180** These ducts provide the <u>route</u> by which the sperm cells may leave the male body on their way to the female body and are thus _____ organs of the male.
accessory	**8-181** On their way through the ducts, the sperm cells are joined by fluid substances secreted by other accessory organs, notably the PROSTATE (PROSS-tate) GLAND and the SEMINAL VESICLES (VES-ik-ls). The sperm cells will need the secretions of the _____ gland and the _____ vesicles as a medium in which to travel to and in the female body.
prostate seminal	**8-182** Sperm cells alone cannot reach the ovum without the fluid medium provided by the secretions of the _____ gland and the _____ _____.
prostate seminal vesicles	**8-183** The substance which is formed by the combination of the sperm cells and the secretions of the prostate gland and the seminal vesicles is called SEMEN (SEE-men). Since the sperm cells cannot reach the ovum without their fluid medium, it is as components of this substance, _____, that they make the rest of the trip.
semen	**8-184** The sperm cells leave the male body and enter the female body, during copulation, through the PENIS (PEE-nis). The copulatory organ of the male reproductive system is the _____, which provides the means by which the sperm cells--contained in the substance called _____ -- can enter the female body.
penis semen	**8-185** The final accessory organ of the male reproductive system is the copulatory organ, the _____.
penis	**8-186** Thus, the principle organs of the male reproductive system are the primary organs, the (1) _____ ,and the accessory organs, including the (2) _____ through which the sperm travels; the (3) _____ gland and the (4) _____ _____ ,which provide the fluid medium for the sperm cells; and the copulatory organ, the (5) _____.

1. testes
2. ducts
3. prostate
4. seminal
   vesicles
5. penis

8-187　　This drawing shows the locations of the principle organs of the male reproductive system.

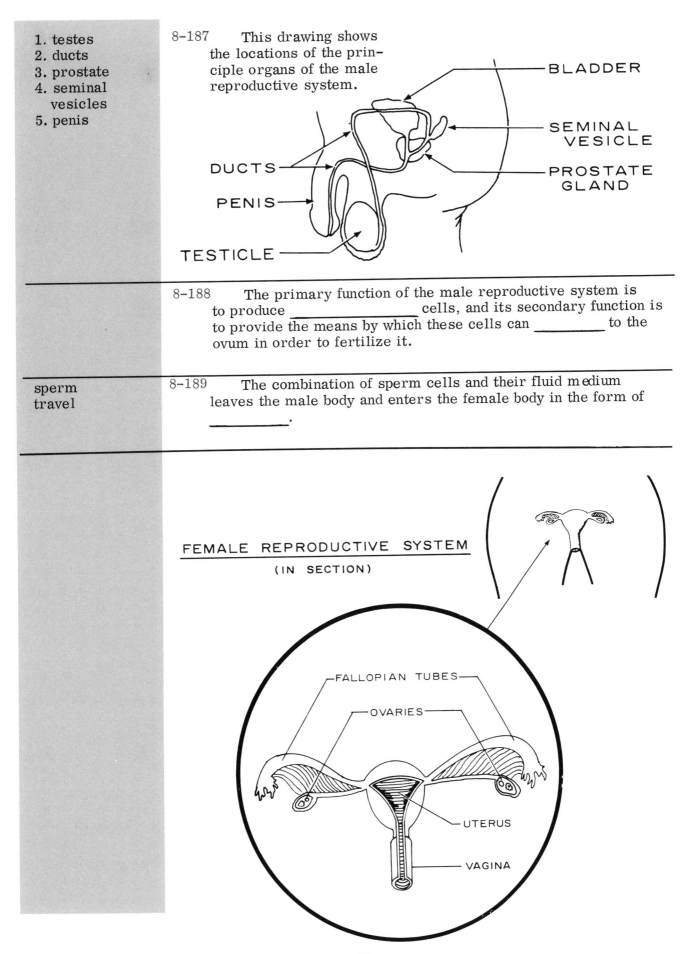

BLADDER

SEMINAL VESICLE

PROSTATE GLAND

DUCTS

PENIS

TESTICLE

8-188　　The primary function of the male reproductive system is to produce _____ cells, and its secondary function is to provide the means by which these cells can _____ to the ovum in order to fertilize it.

sperm
travel

8-189　　The combination of sperm cells and their fluid medium leaves the male body and enters the female body in the form of _____.

FEMALE REPRODUCTIVE SYSTEM
(IN SECTION)

FALLOPIAN TUBES

OVARIES

UTERUS

VAGINA

293

semen	8-190 **THE FEMALE REPRODUCTIVE SYSTEM**
	Now refer to the drawing of the female system. The part of the female reproductive system which leads to the outside of the body is the tubular channel called the _____.
VAGINA (va-JY-nah)	8-191 The sperm cells, contained in semen, enter the female body through the _____.
vagina	8-192 The vagina leads into the hollow pear-shaped organ called the UTERUS (YOU-ter-us). Sperm cells pass through the vagina and into the _____.
uterus	8-193 Look at the drawing again. Extending from the top of the uterus toward the two OVARIES (the organs which produce the ova) are two tubes called the FALLOPIAN (fal-LOPE-i-an) TUBES. Each ovary is located near the fringed funnel-shaped opening of one of the _____ tubes.
fallopian	8-194 The ovum is produced by the ovary and expelled into the funnel-shaped opening of the _____ _____.
fallopian tubes	8-195 The sperm cells leave the uterus and travel up through the fallopian tubes; the ovum meanwhile travels down the tube until it meets the sperm, and _____ takes place.
fertilization or conception	8-196 Thus, before the sperm cells can fertilize the ovum, they must travel much of the length of the female reproductive system, entering through the (1) _____, passing through the (2) _____, and traveling up the (3) _____ _____ until they meet the ovum produced by the (4) _____.

1. vagina 2. uterus 3. fallopian tube (s) 4. ovary (ies)	8-197 Label the reproductive organs of the female in the drawing below. (1)_____ (2)_____ (3)_____ (4)_____
1. fallopian tube 2. ovary 3. uterus 4. vagina	8-198 Like the testes, which produce sperm, the ovaries, which produce ova, are the _____ organs of the female reproductive system.
primary	8-199 The other functions of the female system, however, are of course completely different from those of the male system. The difference appears first of all in the way sperm and ova are produced, which will be discussed in the following frames. (No answer required.)
	8-200 The male and female gonads both begin producing their respective cells during the period of human growth called puberty. The beginning of sperm or ova production, in fact, is commonly used as the way of defining the onset of _____ in the male or female.
puberty	8-201 Sperm production, once it has begun, is a continuous process; in other words, from the time the male reaches _____, his testes are secreting sperm cells in a more or less _____ flow.
puberty continuous	8-202 The ovaries, on the other hand, produce ova only at regular intervals, over a definite period of time or cycle. Thus, ova production, unlike sperm production, is a (continuous/cyclical) _____ process.

cyclical	8-203 The period of time during which ova are produced is called the MENSTRUAL CYCLE. As you probably know, the length of the average _____ cycle is about 28 days.
menstrual	8-204 An ovary expels an ovum once every 28 days (on an average) -- in other words, once during every _____ cycle.
menstrual	8-205 The menstrual cycle, however, involves more than just the production of ova. Every time an ovum is expelled from an ovary, the organs of the female system must prepare for the possibility that the ovum will be _____ by sperm cells.
fertilized	8-206 Every time the newly-produced ovum <u>fails</u> to be fertilized, these preparations are abandoned, to begin all over again during the next _____ _____.
menstrual cycle	8-207 This process of producing an ovum, preparing for fertilization and the ensuing pregnancy, and abandoning these preparations when fertilization does not occur takes place in the female reproductive organs approximately every _____ days beginning at puberty.
28	8-208 The menstrual cycle is suspended only when fertilization does occur: during the ensuing _____ and for (generally) a short time afterwards.
pregnancy	8-209 The following frames will examine in more detail the way in which the female reproductive organs (1) _____ ova and (2) _____ for the possibility that the ovum will be (3) _____ by a sperm cell.

1. produce 2. prepare 3. fertilized	8-210	In the average 28-day menstrual cycle, the expulsion of the ovum from the follicle of the ovary takes place on the 14th day. The expulsion of the ovum thus marks the (middle/end) _____ of the cycle.
middle	8-211	The first half of the cycle is taken up with three important processes which (precede/follow) _____ the expulsion of the ovum.
precede	8-212	The first of these is the process by which the follicle-stimulating hormone of the pituitary (which you encountered in Part 1 of this chapter) stimulates the _____ to grow to maturity. An ovum can be expelled only by a _____ follicle.
follicle mature	8-213	The follicle requires about 14 days of pituitary stimulation to grow to _____.
maturity	8-214	During this same period, the follicle-stimulating hormone also acts to produce an increased secretion of estrogen from the ovary. As the first half of the cycle progresses, more and more _____ is secreted by the ovary.
estrogen	8-215	The estrogen secreted by the ovary in turn stimulates an increased growth of the cells in the lining of the uterus or womb-- thus preparing for the possibility that the ovum will be fertilized. If the ovum is fertilized, it will lodge ultimately in the _____ of the uterus to continue its development.
lining	8-216	The second process taking place in the first half of the menstrual cycle leading up to the expulsion of the ovum is the preparation of the _____ of the _____ or womb for the possibility of pregnancy.

lining uterus	8-217 The third process which must take place before an ovum is expelled is the formation of the ovum itself from its <u>parent</u> cell in the follicle. The follicle begins each cycle as a cluster of follicle cells enclosing a single _____ cell.
parent	8-218 As the first half of the cycle progresses, the follicle cells and this enclosed _____ _____ mature simultaneously.
parent cell	8-219 About 2 or 3 days before the middle of the cycle, the parent cell begins to divide. Reproductive cell division, however, is quite different from the process of _____ undergone by other cells.
division	8-220 The ovum which results from the division of the parent cell contains only <u>half</u> the usual number of chromosomes. When the ovum is ready for expulsion from the follicle, in other words, it contains only _____ (number) chromosomes.
23	8-221 The ovum contains only 23 chromosomes because it is only one of the two parties to the new cell which will be formed if fertilization occurs. The sperm cells, as you may guess, have gone through a similar process of _____ and also have only 23 _____.
division chromosomes	8-222 The cell which will be formed if the ovum unites with a sperm will thus have the full complement of _____ (number) human chromosomes.
46	8-223 The process of division undergone by the parent cell in the follicle is therefore another way in which the female reproductive organs are _____ for the possibility of fertilization.

preparing	8-224 By the 14th day of the cycle, the follicle is fully (1) _____, and the (2) _____ of the uterus is developed. The ovum-- containing (3) _____ chromosomes -- is expelled from the follicle into the funnel-shaped end of the (4) _____ tube.
1. matured 2. lining 3. 23 4. fallopian	8-225 If the ovum is not fertilized after it is expelled from the follicle, it disintegrates in the fallopian tube, and the cycle continues. As you know, the menstrual cycle runs out its 28-day period every time that fertilization (does/does not) _____ occur.
does not	8-226 The estrogen secretion of the ovaries begins to <u>decrease</u> with the expulsion of the ovum. As the second half of the cycle progresses, the cells in the lining of the uterus receive progressively (less/more) _____ estrogen stimulation.
less	8-227 By the 28th or last day of the cycle, the lining begins to deteriorate. The flow of blood through the vagina which punctuates the menstrual cycle is caused by the _____ and subsequent shedding of the lining of the uterus.
deterioration	8-228 You should note that the onset of bleeding is considered to mark the <u>beginning</u> of the menstrual cycle; in other words, the bleeding period occupies the (first/last) _____ few days of the cycle.
first	8-229 After an average bleeding period of about 4 days, the follicle begins to mature again; the production of estrogen by the ovaries increases, and a new menstrual _____ is under way.
cycle	8-230 The following frames will examine what happens when fertilization <u>does</u> occur, and the menstrual cycles are suspended during the ensuing pregnancy. (No answer required)

If sperm is present in the fallopian tube, the ovum will probably be fertilized. The newly fertilized ovum travels slowly down the tube. By about the tenth day after it was expelled from the follicle, it has become attached to the _____ of the uterus.

---

lining

8-232   As estrogen secretion diminishes, the pituitary stimulates the ovaries to produce the hormone progesterone (pro-JEST-er-own), which was mentioned in the first part of this unit. The job of maintaining the lining of the uterus is largely taken over from estrogen by _____.

---

progesterone

8-233   As estrogen is the primary sex hormone of the female, so _____ is the secondary sex hormone.

---

progesterone

8-234   Progesterone is secreted by the follicle from which the ovum was expelled. A follicle which has expelled its ovum is yellow in color and is called a CORPUS LUTEUM (LU-tee-um), "corpus" meaning "body" and "_____" meaning "yellow."

---

luteum

8-235   By stimulating the continued growth of the cells in the lining of the uterus where the fertilized ovum is housed, the _____ _____ , which produces progesterone, plays an important role in early pregnancy.

---

corpus luteum

8-236   The pituitary hormone primarily responsible for stimulating the corpus luteum to produce progesterone is the LUTEINIZING (LOO-tee-in-izing) HORMONE, abbreviated LH. What the follicle-stimulating hormone (FSH) is to estrogen, the _____ _____ (LH) is to progesterone.

luteinizing hormone	8-237	Fill in the blanks below:

8-237   Fill in the blanks below:

PITUITARY

_____ _____ hormone _____ hormone

    MATURING   CORPUS
    FOLLICLE   LUTEUM

estrogen   progesterone
             UTERUS

---

follicle-stimulating / **luteinizing**

8-238   You should note, however, that although the luteinizing hormone stimulates a sharp rise in the production of progesterone after the expulsion of the ovum, the production of estrogen by no means ceases; and the levels of both progesterone and estrogen remain high throughout pregnancy.

8-239   The early stages of pregnancy are marked by the development of two organs essential to the growth of the child in the uterus. These organs, as well as the child itself, develop from the fertilized _____.

---

ovum

8-240   After the fertilized ovum has become imbedded in the lining of the uterus, a part of it develops into the pancake-like organ called the PLACENTA (pla-SENT-ah). It is through this organ, the _____, that the developing child will receive its nourishment.

---

placenta

8-241   The placenta absorbs <u>nutrients</u> from the mother's blood stream and transmits them through the membranous UMBILICAL (um-BIH-li-kl) CORD to the developing child. At the same time, waste materials are returned from the child through the _____ cord to the _____ for disposal through the mother's blood stream.

---

umbilical / **placenta**

8-242   The blood supply of the developing child is shared by the _____,which absorbs the child's nutrients from the mother, and the _____ _____ through which these nutrients pass.

placenta umbilical cord	**8-243**　　It is important to note, however, that the blood supply of the <u>mother</u> is <u>completely</u> <u>separate</u> from the blood supply shared by the child, the umbilical cord, and the placenta. Although nutrients and waste materials pass across the placenta, the blood of the mother and the blood in the placenta (do/do not) _____ intermix.
do not	**8-244**　　By the time the placenta and the umbilical cord have been formed (during the second month of pregnancy), all the basic mechanisms are established which will support the developing child until _____ -- approximately 9 months after conception.
birth	**8-245**　　The processes by which a fully-formed child is developed in the uterus from a single fertilized ovum are too complex to be covered in the present course. You should, however, be familiar with the terminology which applies to the <u>stages</u> of the child's _____ within the uterus.
development	**8-246**　　The first stage of pregnancy is called the <u>period of the ovum</u>. The period of the ovum begins when the ovum is fertilized by the sperm cell and ends when it becomes embedded in the lining of the uterus. Thus the period of the _____ covers about the first 10 days of pregnancy.
ovum	**8-247**　　The period of the ovum is followed by the period of the EMBRYO (EM-bree-o), which lasts until the end of the third month of pregnancy. The fertilized ovum is called an (1) _____ from the time it becomes (2) _____ in the lining of the uterus until the end of the (3) _____ month in the uterus.
1. embryo 2. embedded 3. third	**8-248**　　By the end of the third month, the placenta and the umbilical cord are fully formed, and most of the organs of the child have begun to take recognizable shape. These developments take place during the period of the _____ , the second stage of pregnancy.
embryo	**8-249**　　The expulsion of an embryo from the uterus for any reason during this period is called an EMBRYONIC ABORTION. The term _____ _____ applies to expulsion of the developing child by any cause, whether spontaneous or induced, during the period of the embryo , that is, up to the end of the _____ month of pregnancy.

embryonic abortion third	8-250	The third and final stage of pregnancy is the period of the FETUS (FEE-tus). From the beginning of the fourth month until birth, the developing child is called a _____.
fetus	8-251	After about seven full months in the uterus, the developing child becomes <u>viable</u>, or capable of surviving expulsion from the uterus. The developing child—called a _____ during this period -- is _____ during the last two months of the normal term of pregnancy.
fetus viable	8-252	The term "miscarriage" means the expulsion of the developing fetus from the uterus before it is capable of surviving, that is, before it becomes _____.
viable	8-253	Miscarriage, however, is a lay term; in medicine the term <u>abortion</u> is used also for expulsion of the <u>fetus</u> before it becomes viable, before the end of the _____ month of pregnancy.
7th	8-254	Thus the three stages of pregnancy are: the period of the (1) _____, about the first (2) _____ days after conception; the period of the (3) _____, up to the end of the (4) _____ month; and the period of the (5) _____.
1. ovum 2. 10 3. embryo 4. third 5. fetus	8-255	The term abortion refers to the expulsion of either the embryo (embryonic abortion) or the fetus so long as it is non-viable, that is, before it becomes _____ _____ _____. (Your own words)
capable of sur- viving expulsion from the uterus (or equivalent statement)	8-256	With the delivery of a viable or live child, birth - the most important business of the reproductive process is completed. The female reproductive organs will go through a short period of rest and readjustment to be ready to begin the process again. Before you begin the section on the urinary system, you may take a break.

As the organs of the reproductive system perform the life function of reproduction, the organs of the urinary system perform the life function of excretion. The urinary organs, as their name indicates, excrete the waste materials of metabolism from the body in the form of _____.

---

urine

8-258   Approximately 75% of the excretory function of the body is is performed by the organs of the _____ system.

---

urinary

8-259   The waste products of metabolism can be harmful and even fatal if allowed to accumulate in the body. Thus the proper functioning of the urinary system, which plays the largest role in removing these materials, (is/is not) _____ essential to the health of the body.

URINARY SYSTEM

(INFERIOR VENA CAVA) — (AORTA)

(RENAL ARTERY) — (ADRENAL GLAND)

KIDNEY

(RENAL VEIN) —

URETER

BLADDER

URETHRA

is	8-260	Look at the drawing of the urinary system, in which the names of the principal urinary organs are underlined.

The most important of these organs are the two KIDNEYS, which are responsible for the <u>formation</u> of urine. Urine is formed by the _____ and passes into two narrow tubes called _____.

kidneys URETERS (YOOR-et-ers)	8-261	Look at the drawing.

Urine passes from the kidneys, where it is formed, through the tubes called _____ and into the _____, where it gradually accumulates.

1. ureters 2. bladder	8-262	As the bladder fills with urine, its muscular walls distend until the point is reached--usually with about a pint of urine-- when the sensory nerves in the walls alert the brain that the _____ is full.

bladder	8-263	Look at the drawing again.

When the bladder is full, urine is discharged from the body through the _____.

urethra (yoo-REE-thra)	8-264	In the male, both urine and semen leave the body by the same route: through the _____ of the penis.

urethra	8-265	In the female body, on the other hand, the opening of the urethra is anterior to that of the vagina and quite separate from it; the reproductive and the urinary systems of the female (do/ do not) _____ share a common pathway.

UTERUS

BLADDER

URETHRA

VAGINA

do not	**8-266** Look at the drawing of the urinary system again. As you can see, the kidneys are in very close touch with the largest blood vessels in the body. The kidneys produce urine initially from the waste materials in the _____ which circulates through them.
blood	**8-267** A single RENAL ARTERY normally supplies all the blood to a single kidney. Once inside the kidney, however, a vast system of arterioles and capillaries branch and rebranch from this one _____ artery.
renal	**8-268** The arterioles and capillaries within the kidney form into duct systems called NEPHRONS (NEF-rons), which are the <u>functional</u> units of the kidneys. The kidney's function of producing urine is performed by the _____.
nephrons	**8-269** Each kidney contains about one million of these functional units or _____.
nephrons	**8-270** The capillaries of the nephron form together into a cluster called the _____.

NEPHRON
(IN PARTIAL SECTION)

AFFERENT ARTERIOLE   EFFERENT ARTERIOLE

GLOMERULUS (CLUSTER OF CAPILLARIES)

BOWMAN'S CAPSULE

RENAL TUBULE |

GLOMERULUS (glo-MARE-you-lus)	8-271 Blood is carried into the glomerulus through an <u>afferent</u> arteriole. Blood is carried <u>out</u> of the _____ through an _____ arteriole.

NEPHRON (IN PARTIAL SECTION)

AFFERENT ARTERIOLE — EFFERENT ARTERIOLE — GLOMERULUS (CLUSTER OF CAPILLARIES) — <u>BOWMAN'S CAPSULE</u> — <u>RENAL TUBULE</u> |
| glomerulus efferent | 8-272 Each glomerulus is enclosed by a funnel-like structure called the _____ _____. |
| Bowman's capsule | 8-273 A tubule called the _____ tubule leads out of the _____ _____, which encloses the glomerulus.

NEPHRON (IN PARTIAL SECTION)

AFFERENT ARTERIOLE — EFFERENT ARTERIOLE — GLOMERULUS (CLUSTER OF CAPILLARIES) — <u>BOWMAN'S CAPSULE</u> — <u>RENAL TUBULE</u> |

renal

Bowman's capsule

8-274   Fill in the blanks in the drawing below.

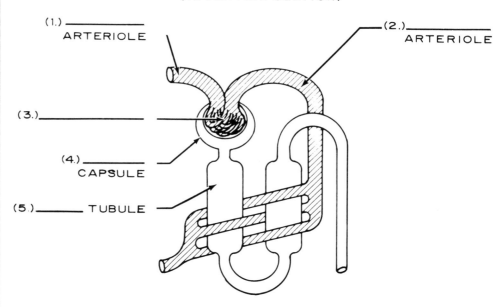

NEPHRON
(IN PARTIAL SECTION)

(1.)_____ ARTERIOLE

(2.)_____ ARTERIOLE

(3.)_____

(4.)_____ CAPSULE

(5.)_____ TUBULE

1. afferent
2. efferent
3. glomerulus
4. Bowman's
5. renal

8-275   The efferent arteriole, which carries the blood out of the
_____, branches again into capillaries which are
elaborately intertwined with the renal tubule which leaves the
_____ _____.

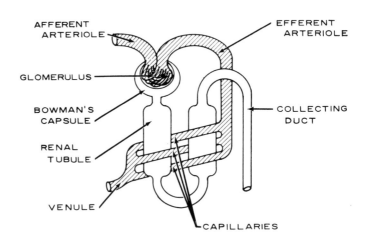

NEPHRON
(IN PARTIAL SECTION)

AFFERENT
ARTERIOLE

EFFERENT
ARTERIOLE

GLOMERULUS

BOWMAN'S
CAPSULE

COLLECTING
DUCT

RENAL
TUBULE

VENULE

CAPILLARIES

glomerulus Bowman's capsule	8-276	The capillaries from the efferent arteriole that twine around the renal tubule empty ultimately into a single _____.

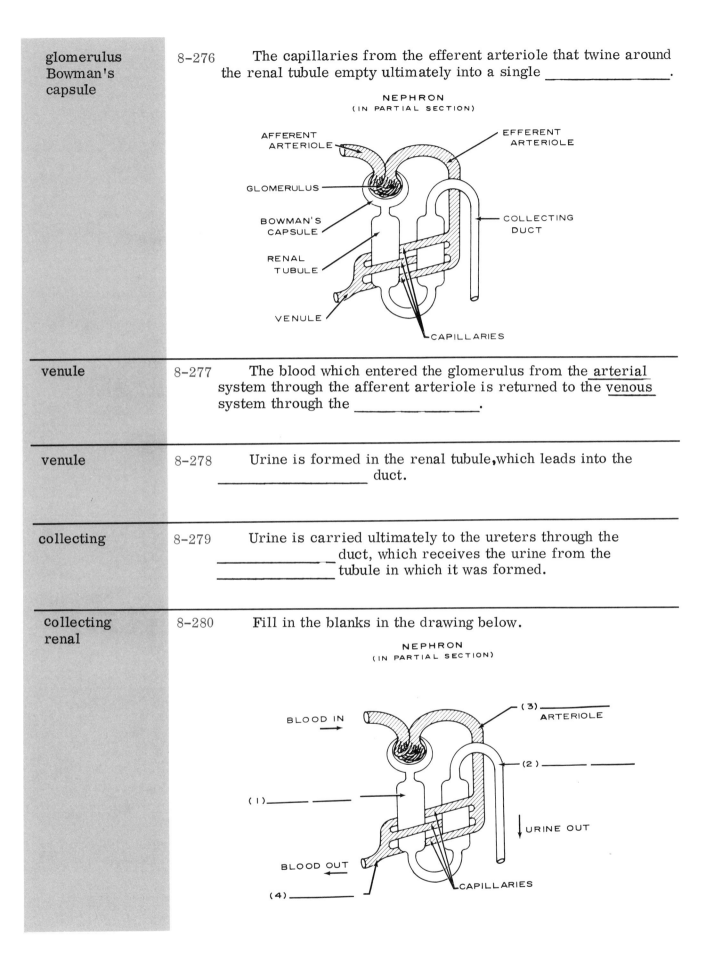

NEPHRON
(IN PARTIAL SECTION)

AFFERENT ARTERIOLE

EFFERENT ARTERIOLE

GLOMERULUS

BOWMAN'S CAPSULE

COLLECTING DUCT

RENAL TUBULE

VENULE

CAPILLARIES

venule	8-277	The blood which entered the glomerulus from the <u>arterial</u> system through the afferent arteriole is returned to the <u>venous</u> system through the _____.
venule	8-278	Urine is formed in the renal tubule, which leads into the _____ duct.
collecting	8-279	Urine is carried ultimately to the ureters through the _____ duct, which receives the urine from the _____ tubule in which it was formed.
collecting renal	8-280	Fill in the blanks in the drawing below.

NEPHRON
(IN PARTIAL SECTION)

BLOOD IN

(3) _____ ARTERIOLE

(2) _____

(1) _____

URINE OUT

BLOOD OUT

(4) _____

CAPILLARIES

309

1. renal tubule 2. collecting duct 3. efferent 4. venule	8-281　　The functional units of the kidney are the _____ , of which each kidney contains about _____ _____ (number).
nephrons one million	8-282　　Urine is formed, within the nephron, in the renal (1)_____ , which begins at a funnel-like structure called the (2) _____ _____ and leads out of the nephron into a (3) _____ _____ .
1. tubule 2. Bowman's capsule 3. collecting duct	8-283　　Blood enters the glomerulus from the arterial system through an afferent (1) _____ ,which carries it into the cluster of capillaries called the (2) _____ . Blood passes out of this cluster again into an (3)_____ _____ , which branches again into capillaries that intertwine with the renal tubule. Blood leaves the nephron for return to the venous system through a (4) _____ .
1. arteriole 2. glomerulus 3. efferent arteriole 4. venule	8-284　　The following frames will cover the processes involved in the production of urine. 　　(No answer required).
	8-285　　The first stage of urine production is the <u>filtration</u> of blood by the glomerulus. As blood flows through the capillaries of the glomerulus, about 20 per cent of it is _____ through the walls of the capillaries and into the Bowman's _____ .
filtered capsule	8-286　　Two of the normal constituents of whole blood are too large to pass through the walls of the capillaries: particles of serum protein and the blood cells. The substance which is filtered into the Bowman's capsule, then, is simply blood minus its particles of serum _____ and its _____ .
protein cells	8-287　　The substance filtered into the Bowman's capsule by the glomerulus is called the <u>glomerular filtrate</u>. The glomerular filtrate represents about 20 per cent of the volume of blood passing through the glomerulus, and consists of blood minus _____ _____ and blood _____ .

serum protein cells	8-288 The force required to produce the glomerular filtrate is provided by the blood pressure. More or less than 20 per cent of the blood in the glomerulus may be filtered through the capillary walls according to variations in the blood _____ of the individual.
pressure	8-289 The glomerular filtrate is funneled from the Bowman's capsule into the renal _____.
tubule	8-290 The glomerular filtrate is <u>not</u> urine. Whole blood, as you know, contains many substances needed by the body, and all of these (except protein and blood cells) are filtered indiscriminately into the renal tubule. It is essential to the health of the body that these substances (be/not be) _____ allowed to leave the body in urine.
not be	8-291 The process by which substances essential to the body are transferred out of the glomerular filtrate and back into the blood stream is called <u>reabsorption.</u> Electrolytes, glucose, and water are the most important of the substances returned to the blood by the process of _____.
reabsorption	8-292

REABSORPTION

BLOOD IN →

GLOMERULAR FILTRATE —

URINE OUT

BLOOD OUT ←

CAPILLARIES

(No response required)

8-293   "Reabsorption" means that as the glomerular filtrate flows through the renal tubule, the cells in the walls of the tubule itself <u>reabsorb</u> whatever substances are needed and pass them out into the _____ which are twined around the tubule.

capillaries	8-294 Over 99 per cent of the glomerular filtrate is normally _____ into the walls of the tubule and returned to the _____ in the surrounding capillaries.
reabsorbed blood	8-295 The renal tubule, however, is very selective about what it reabsorbs. In a manner not yet fully understood, the tubule is able to _____ for reabsorption just the amount of any one substance actually needed by the blood.
select	8-296 Thus the glomerular filtrate contains the charged ions which make up the <u>electrolyte</u> content of blood. The renal tubule is able to determine whether there is a healthy balance between the positively and negatively charged _____ in the filtrate.
ions	8-297 If there are too many ions of either charge, the renal tubule will not reabsorb the excess ions, leaving them to be excreted in the urine. By thus returning a balanced proportion of electrolytes to the blood stream, the renal tubule plays a vital role in maintaining the _____ balance of the whole body.
electrolyte	8-298 Besides electrolytes, another substance selectively reabsorbed by the tubule is <u>glucose</u> or sugar. When the glucose level in the filtrate is within normal limits, the walls of the tubule reabsorb <u>all</u> of it; in other words, <u>no</u> _____ or sugar is excreted.
glucose	8-299 Glucose is excreted in urine only when the level of glucose in the filtrate rises above healthy limits, and the excess (is/is not) _____ reabsorbed by the tubule.
is not	8-300 Thus the renal tubule regulates not only the _____ balance but also the level of _____ _____ in the blood.

electrolyte glucose sugar	8-301 A third substance reabsorbed by the renal tubule is water. Of the total volume of water in the filtrate, the tubule normally returns about 85% to the capillaries in a passive way and exercises its selective faculty over the remaining _____ %.
15	8-302 How much of this remaining 15% will be reabsorbed by the tubule is determined by a posterior pituitary hormone called the ANTI-DIURETIC (die-you-RET-ik) HORMONE. The posterior pituitary is stimulated to produce _____ - _____ hormone by changes in the body's water balance.
anti-diuretic	8-303 "Diuretic" means "tending to increase the volume of urine excreted." The volume of urine is made up almost entirely of water. Thus the anti-diuretic hormone would act to (increase/ decrease) _____ the amount of water left to be excreted in urine.
decrease	8-304 When the level of water in the body falls below normal -- for instance, through excessive sweating -- the renal tubule is stimulated to reabsorb whatever amount of water is needed by an increased production of _____ - _____ hormone from the posterior pituitary.
anti-diuretic	8-305 Similarly, when the body takes on more than the usual amount of water, the pituitary will produce (more/less) _____ anti-diuretic hormone, and the renal tubule will reabsorb (more/less) _____ water for return to the blood.
less less	8-306 Anti-diuretic hormone is commonly referred to by its initials, A.D.H. You should remember that A.D.H. acts to increase reabsorption of only the _____ % of water in the filtrate not normally reabsorbed by the tubule as a matter of course.
15	8-307 Another factor besides A.D.H. which influences water reabsorption is the RENAL SOLUTE LOAD: the load of _____ or substances dissolved in the renal tubule.

solute (s)	8-308	The load or concentration of solutes filtered into the renal tubule influences water reabsorption through its effect on the process of <u>osmosis</u>, which you learned in an earlier chapter as the tendency of your body fluids to pass through semi-permeable membranes from the side with the (lower/higher) _____ concentration of solutes to the side with the (lower/higher) _____ concentration.
lower higher	8-309	The passive reabsorption of the first 85% of water takes place by osmosis; that is, as the solutes (glucose, ions, etc.) initially filtered into the renal tubule are selectively reabsorbed, their concentration rises inside the _____ surrounding the renal tubule, and the water simply follows them.
capillaries	8-310	However, if there is an abnormally high renal solute load-many more solutes in the filtrate than are reabsorbed - the concentration in the tubule may be such that an excessive amount of water (passes out of/remains in) _____ the tubule and is subsequently secreted.
remains in	8-311	Thus the effect of an abnormally high renal _____ load is typically diuretic: it tends to _____ the amount of urine excreted.
solute increase	8-312	The diuretic effect of an abnormally high _____ _____ _____ may be harmful if it results in the loss of _____ which should otherwise have been retained in the body.
renal solute load water	8-313	The third part of urine production is <u>secretion</u>: substances are transferred out of the _____ in the capillaries and into the cells of the tubule, which then _____ these substances into the filtrate.
blood secrete	8-314	The substances secreted into the filtrate are foreign elements in the blood which were not filtered into the Bowman's capsule by the glomerulus. These _____ substances are excreted from the body in urine.

8-315   The formation of urine is now completed. Blood has been (1)_____ through the glomerulus into the Bowman's capsule, which passes it into the renal tubule. Substances essential to the body have been (2) _____ by the tubule and returned to the capillaries entwined around the tubule.

    Foreign elements have been (3) _____ into the filtrate.

---

1. filtered
2. reabsorbed
3. secreted

8-316   The water volume of the glomerular filtrate has been reduced first by passive (osmotic) reabsorption and then by the action of A.D.H. until the end product is highly concentrated; this end product is the urine, which passes through and eventually leaves the nephron, the kidney, the bladder and finally the _____.

---

body

8-317   This is the end of the chapter on the Genito-Urinary System.

CHAPTER

9

The Digestive System

9-1   The DIGESTIVE SYSTEM is essentially a long tube extending from the opening of the mouth at one end to the opening of the rectum --the anus -- at the other end.

The major parts of the digestive system are the <u>mouth</u>, the <u>pharynx</u>, the <u>esophagus</u>, the <u>stomach</u>, the <u>intestines</u>, and the associated organs: the <u>liver</u>, <u>gall bladder</u> and <u>pancreas.</u>

9-2   The function of the digestive system is to break down foodstuffs -- carbohydrates, fats, and proteins -- into molecules which can be absorbed by the blood stream and carried to the cells of the body.

You will investigate the digestive system by tracing the breakdown of foodstuffs in their course from the mouth to their eventual entry into the _____stream and hence their arrival at the

_____ of the body.

blood
cells

9-3   The foodstuffs broken down by the digestive system can be divided into two general classes: (1) substances that serve as direct sources of energy and (2) substances that are necessary for life activities but that do not serve as direct sources of

_____.

energy

9-4   Vitamins, minerals and water are necessary for life activities but <u>do not</u> provide energy directly. Carbohydrates, fats, and proteins, however, do serve as immediate sources of

_____.

energy	9-5	The foodstuffs that provide energy when broken down by the digestive system are: a. carbohydrates, fats and proteins. b. vitamins, minerals and water.
a. carbohy- drates, fats and proteins	9-6	The foodstuffs that support life activities but do not provide energy directly when broken down by the digestive system are: a. carbohydrates, fats and proteins. b. vitamins, minerals and water.
b. vitamins, minerals and water	9-7	The process of digestion involves the breakdown of food-stuffs -- both energy-providing and non energy-providing -- into molecules which are absorbed into the _____ stream and are carried to all the _____ of the body.
blood cells	9-8	The Mouth, Pharynx and Esophagus Before digestion begins, food is ingested--taken into the mouth--and the tongue, which is an accessory organ, moves it into a position in which it is ground into smaller pieces by the _____ , another accessory organ.
teeth	9-9	Chewing helps to prepare the food for swallowing but is not an integral or necessary part of the digestive process. The digestive process starts when the SALIVARY GLANDS, another accessory organ, secrete _____,which moistens the food.
saliva	9-10	Saliva, by moistening the food, facilitates swallowing. In addition, saliva begins the process of digestion--food breakdown-- by attacking the carbohydrates. The first of the foodstuffs to be broken down, the _____, are in the energy-producing class of foods.
carbohydrates	9-11	The breakdown of the carbohydrates is started in the mouth by PTYALIN (TI-a-lin). One of the components of saliva is an enzyme (an organic substance that accelerates chemical trans-formation) which attacks carbohydrates. This enzyme is called _____.

ptyalin

9-12   Saliva facilitates swallowing and begins the digestion of the carbohydrates in the food. Molecules of starch (carbohydrates) are split into smaller molecules of sugar (carbohydrates) by the enzyme called _____.

THE DIGESTIVE SYSTEM

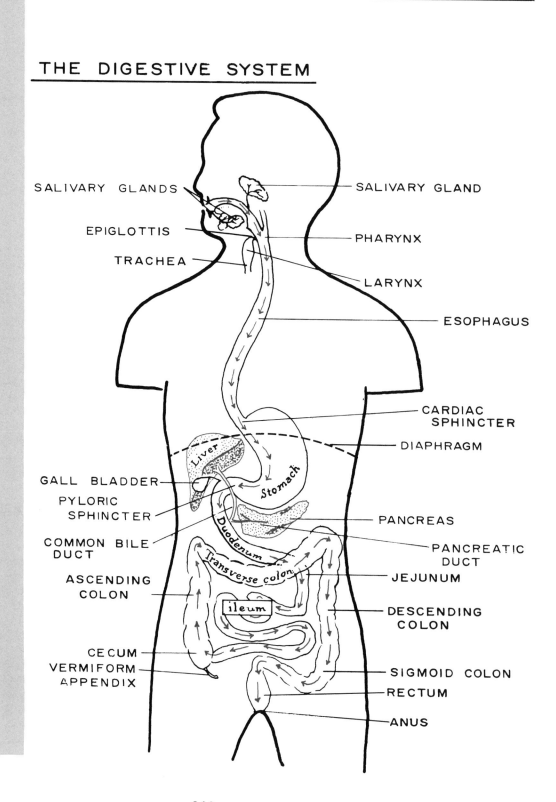

SALIVARY GLANDS

SALIVARY GLAND

EPIGLOTTIS

PHARYNX

TRACHEA

LARYNX

ESOPHAGUS

CARDIAC SPHINCTER

DIAPHRAGM

Liver

Stomach

GALL BLADDER

PYLORIC SPHINCTER

Duodenum

PANCREAS

COMMON BILE DUCT

PANCREATIC DUCT

Transverse colon

JEJUNUM

ASCENDING COLON

ileum

DESCENDING COLON

CECUM

VERMIFORM APPENDIX

SIGMOID COLON

RECTUM

ANUS

ptyalin	9-13	<u>Turn to the diagram of the digestive system and refer to it as needed.</u> A pulpy mass of food that has been mixed with saliva is called a BOLUS (BO-lus). The next step in the digestive process is the movement of the bolus by the tongue into the _____, which is a common passageway for both food and air.
<u>PHARYNX</u> (FAR-inks)	9-14	While food remains in the mouth, the act of swallowing the food is voluntary; one can either begin to swallow or not, as one chooses. When food reaches the pharynx, however, swallowing ceases to be a voluntary act and becomes an automatic, or _____ act or reflex.
involuntary	9-15	Swallowing a piece of food is a voluntary act while the food is in the _____. Swallowing becomes an involuntary, or automatic, act when the food reaches the _____.
mouth pharynx	9-16	The soft palate is moved up and back, closing off the respiratory pharynx. This action prevents the food and liquids from coming out of the _____.
nose	9-17	The tongue moves the bolus (the pulpy mass of food and saliva) from the mouth to the pharynx. Then skeletal muscles in the pharynx propel the bolus into the tube leading to the stomach, namely the _____.
ESOPHAGUS (e-SOF-a-gus)	9-18	As you will recall from the unit on respiration, the pharynx opens not only into the esophagus, but also into the _____, or voice box.
larynx	9-19	You will also recall that food is prevented from entering the larynx by a reflex closure of the _____, which is located at the entrance to the larynx.

epiglottis	**9-20** The action of the soft palate prevents food from entering the nasal passage,and the action of the epiglottis prevents food from entering the larynx. These two actions insure that the food will pass directly from the mouth to the _____ and then to the _____.
pharynx esophagus	**9-21** Alternating waves of contraction and relaxation force the bolus down the tube that leads from the pharynx to the stomach. This tube is called the _____.

CONTRACTION
RELAXATION
BOLUS |
| esophagus | **9-22** The esophagus is about ten inches long. Food is forced down the length of the esophagus by alternating waves of _____ and _____ and the force of gravity.

BOLUS

ESOPHAGUS |
| contraction relaxation (either order) | **9-23** This process of alternating contraction and relaxation by which food is pushed down the esophagus is called PERISTALSIS (per-i-STAL-sis) or PERISTALTIC ACTION. In lower portions of the digestive tract, the small intestine for example, food is also moved by a similar _____ action. |
| peristaltic | **9-24** The process by which food is pushed down through the esophagus is called (1) _____ and involves the alternating (2) _____ and (3) _____ of the muscles of the esophagus. |

1. peristalsis (peristaltic action) 2. contraction 3. relaxation ((2) and (3) in either order)	9-25	In the process of swallowing, food is moved from the mouth to the (1) _____. An involuntary action then projects the food into the (2) _____. At that point (3) _____ _____ takes over to move the food down into the stomach.
1. pharynx 2. esophagus 3. peristaltic action	9-26	Turn again to the drawing of the digestive system. As you can see, the lower end of the esophagus is made up of an area of muscle called the CARDIAC SPHINCTER (KAR-di-ak SFINCK-ter). Just before the food reaches the stomach, it passes through a region called the _____ _____.
cardiac sphincter	9-27	Food is prevented from refluxing (returning) from the stomach to the esophagus by a region at the lower end of the esophagus. This area is called the _____ sphincter.
cardiac	9-28	The cardiac sphincter is an area of smooth muscle in the wall of the esophagus which serves to _____ _____. (Complete the sentence in your own words).
prevent the reflux of food from the stomach to the esophagus	9-29	You have traced food from the mouth to the stomach where the first major contributions to the process of digestion take place. Before continuing, let's review the steps in the digestive process that you have studied so far. (No response required-move to frame 30).
	9-30	Upon being placed in the mouth, food is chewed. Chewing aids the swallowing process but (1) _____(does/does not) otherwise affect digestion. The first contribution to the digestive process is made by the (2) _____,which starts to break down the (3)_____ in the food.
1. does not 2. saliva 3. carbohydrates	9-31	The salivary enzyme, _____, starts to split complex starches into simple sugars. Once moistened with saliva,the food, now called a bolus, is moved by the tongue to the _____.

ptyalin
pharynx

9-32   Once in the pharynx, the bolus is involuntarily moved into the next part of the digestive tract, the _____.

       The food is then moved downward by a process of alternating contractions and relaxations called _____.

---

esophagus
peristalsis

9-33   Just before entering the stomach, the bolus passes through an area of the esophagus which prevents food from refluxing from the stomach to the esophagus. This region is called the

_____ _____.

---

cardiac
sphincter

9-34   The food has now reached the stomach, an organ that makes a significant contribution to the process of digestion.

       (No response required—move to frame 35).

9-35

THE STOMACH

BOLUS

CARDIA

       The stomach can be divided into four areas. As you can see from the diagram, food first reaches the area called the

_____.

CARDIA (KAR-di-a)	9-36 The cardia surrounds the opening of the esophagus. To the left in the body (but right in the drawing) is a pouch-like portion called the _____.	 THE STOMACH

9-37   The area of the stomach that surrounds the opening of the esophagus is called the _____. The pouch-like portion to the left is called the _____.

FUNDUS
(FUN-dus)

cardia fundus	9-38 The largest portion of the stomach is called the _____.	 THE STOMACH
BODY	9-39 The fourth area of the stomach is the region leading into the intestine. This area is called the _____.	 THE STOMACH

PYLORUS (pi-LOR-us)	9-40	The area leading into the body of the stomach from the esophagus is the _____. The area leading out of the stomach is the _____.

THE STOMACH

cardia pylorus	9-41	The remaining two areas are the pouch-like distention to the left, called the _____, and the main portion of the stomach, called the _____.

THE STOMACH

fundus
body

9-42   Match the names on the left with the descriptions on the right.

1. cardia   a. main portion of stomach

2. body   b. pouch-like area to left

3. pylorus   c. area immediately surrounding the esophagus.

4. fundus   d. area leading to intestines

1. c 2. a 3. d 4. b	9-43 Now let's see how each of these areas of the stomach contributes to the digestive process. Approximately 15 million <u>GASTRIC GLANDS</u> are spread through all of the areas of the stomach. Different types of cells make up these glands, and consequently the _____ _____ differ in various parts of the stomach.
gastric glands	9-44 In the cardia and in the pylorus, the cells are predominantly of the MUCOUS (MU-kus) types. Thus, the secretion of the _____ _____ in these two areas is mainly mucus, a thick, slimy, clear, protective, semi-fluid substance. MUCOUS —> <— HCL and PEPSIN THE STOMACH
gastric glands	9-45 Mucous is predominantly secreted by the gastric glands in the areas that begin and end the stomach, namely in the _____ and in the _____.
cardia pylorus (either order)	9-46 The glands in the <u>fundus</u> and in the <u>body</u> of the stomach are composed mainly of <u>PARIETAL</u> (pa-RI-e-tal) cells and <u>CHIEF</u> cells--in addition to some mucous cells. The parietal cells are responsible for the formation of hydrochloric acid, and the chief cells for the production of the enzyme <u>PEPSIN</u> (PEP-sin). The two main substances secreted by the glands of the fundus and body of the stomach are _____ and _____.
hydrochloric acid pepsin (either order)	9-47 In the fundus and body of the stomach, the gastric glands mainly secrete hydrochloric acid and pepsin, while in the cardia and pylorus the primary secretion is _____.

mucus	9-48	Match each of the four areas of the stomach with the correct secretion.

1. cardia   a. pepsin

2. fundus   b. hydrochloric acid

3. body   c. mucus

4. pylorus

MUCOUS ——→   ←—— HCL and PEPSIN

THE STOMACH

---

1. c

2. a and b

3. a and b

4. c

9-49   In the course of a 24 hour period, the gastric glands secrete approximately 1,000 to 1,500 cc. of GASTRIC JUICE. This gastric juice is composed of the three secretions just mentioned: (1) _____, (2) _____, and (3)_____ _____ plus water and inorganic substances such as sodium, potassium, calcium, sulfate, and phosphate.

---

1. mucus

2. pepsin

3. hydrochloric acid (any order)

9-50   The muscular walls of the stomach contract and relax, churning the food and mixing it with the _____ _____ that is secreted.

---

gastric juice

9-51   Functions of the Gastric Juice

Hydrochloric acid performs an important function--it releases the enzyme pepsin.

Pepsin is secreted as PEPSINOGEN (pep-SIN-o jen). Pepsinogen is not an active enzyme until it is converted to pepsin. This conversion takes place in the presence of _____.

---

hydrochloric acid

9-52   Once pepsin has been formed, it is capable of converting more pepsinogen to pepsin. The whole process is started, however, when pepsinogen is converted to _____ by _____ _____.

pepsin hydrochloric acid	9-53	Thus, the important function of hydrochloric acid is to _____ pepsinogen by converting it to _____.
activate pepsin	9-54	Pepsin is a digestive enzyme. Once it is released by hydrochloric acid (HCl), pepsin acts upon proteins, breaking them down into smaller molecules and AMINO ACIDS (a-me-no), preparing them for absorption into the blood stream. Thus, the digestion of _____ is begun in the stomach.
proteins	9-55	The enzyme ptyalin, which is secreted by the salivary glands, acts on carbohydrates. The enzyme pepsin, which is secreted in the fundus and body of the stomach, breaks _____ down into smaller molecules and _____ acids.
proteins amino	9-56	The third important component of gastric juice is secreted mainly in the cardia and pylorus although it is secreted to a small degree in all parts of the stomach. This component is called _____.
mucus	9-57	If the stomach lining were not protected, it would be attacked by the hydrochloric acid (HCl) and the pepsin in the stomach. Fortunately, the secretion of the glands in the cardia and pylorus is a thick, slimy, protective, semi-fluid substance. This substance, called _____, serves to _____ the stomach from being damaged by the acid and enzymes present.
mucus protect	9-58	To summarize the functions of the various components of gastric juice, first label each of the four sections of the stomach as to the substance secreted.

1.

2.

THE STOMACH

327

1. mucus
2. HCl
    pepsin
	9-59 Now, match each of the secretions listed below with the appropriate function.
	
	     1. mucus a. break proteins into smaller molecule and amino acids.
	
	     2. hydrochloric acid b. line and protect the stomach
	
	     3. pepsin c. activate pepsinogen

1. b 2. c 3. a (and c)	9-60 Before going on to the next stage in the digestive process, let's discuss acidity in the stomach. As you would suspect from its composition, gastric juice is _____. 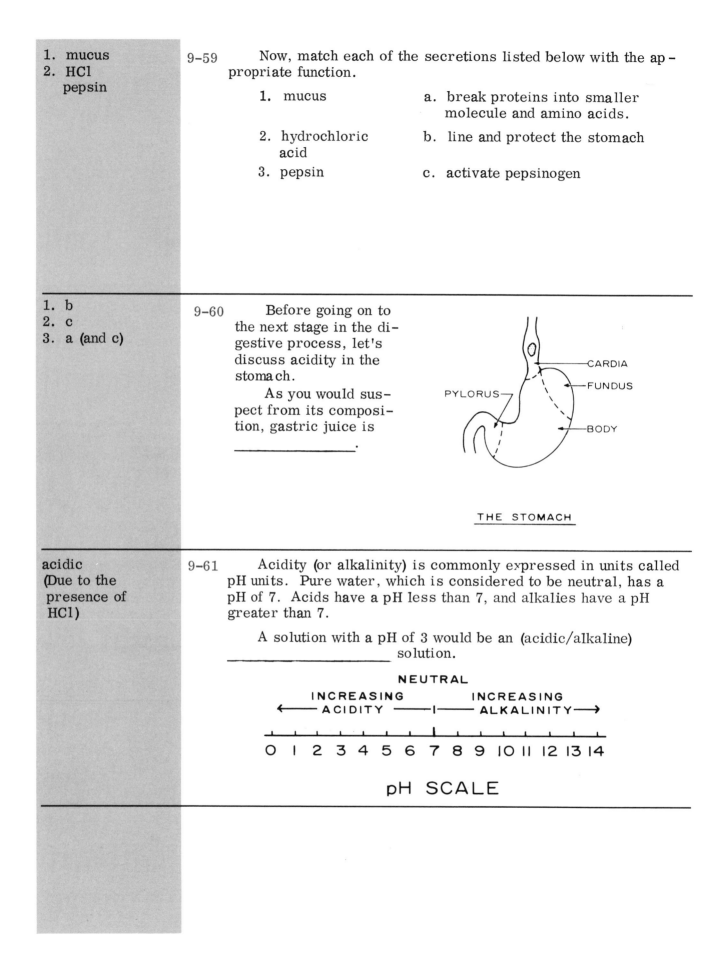 THE STOMACH
acidic (Due to the presence of HCl)	9-61 Acidity (or alkalinity) is commonly expressed in units called pH units. Pure water, which is considered to be neutral, has a pH of 7. Acids have a pH less than 7, and alkalies have a pH greater than 7. A solution with a pH of 3 would be an (acidic/alkaline) _____ solution. NEUTRAL INCREASING INCREASING ← ACIDITY ——|—— ALKALINITY → 0 1 2 3 4 5 6 7 8 9 10 11 12 13 14 pH SCALE

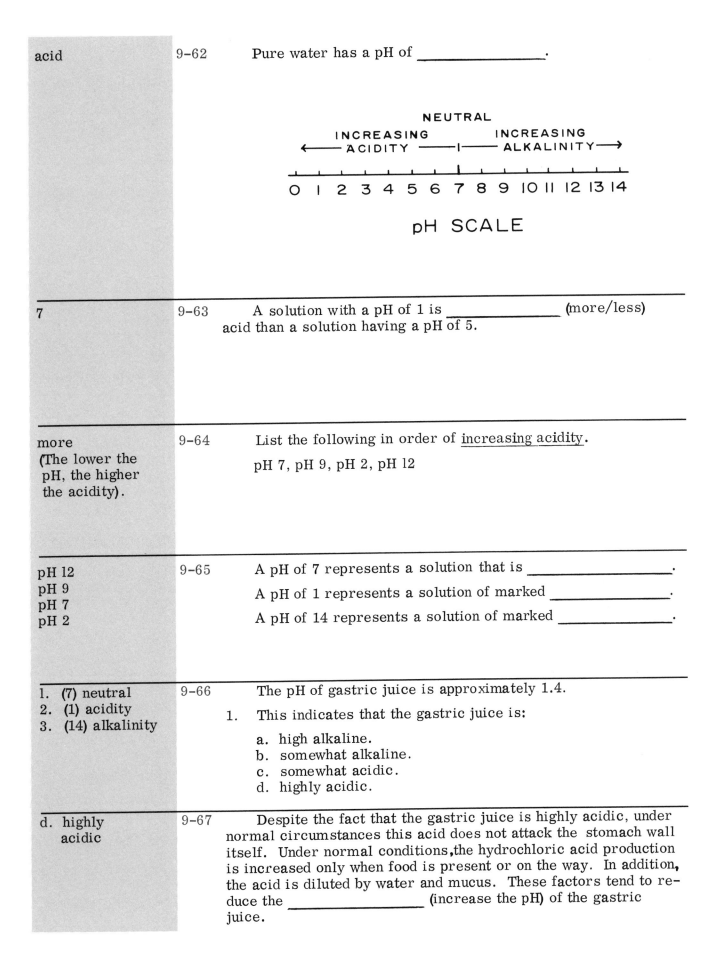

acid	9-62	Pure water has a pH of _____ .

NEUTRAL

INCREASING ACIDITY ←———— | ———→ INCREASING ALKALINITY

0 1 2 3 4 5 6 7 8 9 10 11 12 13 14

pH SCALE

7	9-63	A solution with a pH of 1 is _____ (more/less) acid than a solution having a pH of 5.
more (The lower the pH, the higher the acidity).	9-64	List the following in order of <u>increasing acidity</u>. pH 7, pH 9, pH 2, pH 12
pH 12 pH 9 pH 7 pH 2	9-65	A pH of 7 represents a solution that is _____ . A pH of 1 represents a solution of marked _____ . A pH of 14 represents a solution of marked _____ .

1. (7) neutral
2. (1) acidity
3. (14) alkalinity
	9-66	The pH of gastric juice is approximately 1.4.
		
		1. This indicates that the gastric juice is:
		
		   a. high alkaline.
		   b. somewhat alkaline.
		   c. somewhat acidic.
		   d. highly acidic.

d. highly acidic	9-67	Despite the fact that the gastric juice is highly acidic, under normal circumstances this acid does not attack the stomach wall itself. Under normal conditions, the hydrochloric acid production is increased only when food is present or on the way. In addition, the acid is diluted by water and mucus. These factors tend to reduce the _____ (increase the pH) of the gastric juice.

329

acidity	9-68	<u>Phases</u> <u>of</u> <u>Gastric</u> <u>Secretion</u> The final topic to be discussed before concluding the stomach's role in digestion is the phases of gastric secretion. Gastric secretion <u>increases</u> in response to certain <u>stimuli</u>. Typically,there are three phases of gastric secretion, representing responses to three different types of _____.
stimuli	9-69	The first phase of gastric secretion is the CEPHALIC (se-FAL-ik) PHASE. During this phase,conscious stimuli such as the <u>desire</u> for food, or the <u>sight</u>, <u>smell</u>, or <u>taste</u> of food, can _____ gastric secretion.
increase or stimulate	9-70	Impulses carried from the brain to the stomach by way of the vagus nerve can cause an increase in the secretion of gastric juice. This phase of secretion, called the cephalic phase, is a response to such conscious stimuli as _____ _____ _____. (List as many as you can).
desire for, or taste of, or sight of, or smell of food.	9-71	The gastric juice that is secreted in response to the desire for food or the sight or smell or taste of food during the _____ phase is high in HCl and pepsin content.
cephalic	9-72	The cephalic phase is the first phase of gastric secretion. The second phase is the <u>GASTRIC PHASE</u>. The contact of ingested food with the glands in the pylorus stimulates the second, or _____ phase.
gastric	9-73	During the gastric phase, food reaching the pylorus stimulates the secretion of a hormone called <u>GASTRIN</u>. This hormone travels in the blood stream and stimulates the gastric glands to secrete gastric juice. Thus, in the gastric phase the sequence of stimuli that causes an increased secretion of gastric juice are: 1. food reaching the (1) _____ ; 2. the secretion of the hormone (2) _____ ; 3. the hormone stimulation of the (3) _____ _____.

1. pylorus 2. gastrin 3. gastric glands	9-74	A gastric juice that is high in HCl and pepsin content is secreted during the first or _____ phase. A gastric juice that is lower in pepsin content than that secreted during the first phase is secreted during the second, or _____ phase.
cephalic gastric	9-75	The third phase of gastric secretion is called the INTESTINAL PHASE. The stimulus during this phase is the entrance of certain foods, such as water, proteins, milk, and alcohol, into the small intestine. In response to these foods, the cells of the _____ stimulate the cells of the stomach to secrete gastric juice.
intestine	9-76	When certain foods reach the small intestine, the cells there are stimulated and in turn the gastric glands are stimulated to secrete gastric juice. This third phase of gastric secretion is called the _____ phase.

intestinal	9-77	To summarize--match each of the phases of gastric secretion with the appropriate stimulus (or stimuli).
		
		1. Gastric a. smell or taste of food
		
		2. Cephalic b. food reaching the pylorus
		
		3. Intestinal c. food reaching the intestine
		
		                  d. secretion of the hormone gastrin
		
		                  e. desire for or sight of food

1. b and d 2. a and e 3. c	9-78	To summarize the stomach's role in digestion: Food enters the stomach from the esophagus after passing through the cardiac sphincter. The stomach can be divided into four distinct areas--the cardia, fundus, body, and pylorus. In the cardia and pylorus, the gastric secretion is primarily (1) _____, while in the fundus and body of the stomach the secretion is made up primarily of (2) _____ and (3) _____.
1. mucus 2. hydrochloric acid 3. pepsin (2. and 3. in either order)	9-79	The HCl functions by converting pepsinogen to (1) _____. The function of the enzyme that is activated by HCl is to break (2) _____ down into smaller molecules and amino acids. The secretion of mucus serves to (3) _____ the stomach lining.

1. pepsin 2. proteins 3. protect	9-80	The gastric juice of the stomach has a pH of approximately 1.4 and is thus highly _____. The secretion of HC1 is increased in the response to food and other stimuli. Damage to the stomach lining is avoided by the presence of water and _____.
acidic mucus	9-81	There are three phases of gastric secretion: the first, or (1) _____, phase which is a response to stimuli from the brain when food is seen, smelled, or tasted; the second, or (2) _____, phase which is a response to the secretion of gastrin in the pylorus; and the third, or (3) _____, phase which is a response to certain foods in the small intestine.
1. cephalic 2. gastric 3. intestinal	9-82	<u>The Small Intestine</u> Food entering the stomach has been chewed and mixed with saliva. In this state it is called, as you recall, a bolus. By the time food leaves the stomach it has been changed into a semi-fluid called <u>CHYME</u> (kyme). It is in this state that food moves from the stomach to the beginning of the small _____.
intestine	9-83	You will recall that a bolus is prevented from refluxing from the stomach to the esophagus by the action of the cardiac sphincter. At the other end of the stomach, food, now called _____, is prevented from escaping from the stomach into the small intestine before it has adequately mixed with HC1 and pepsin by the action of the pyloric _____. (Use the diagram of the digestive system as needed)
chyme sphincter	9-84	As the drawing on p. 318 indicates, chyme, the product of the digestive action of the stomach, passes through the _____ _____ into the DUODENUM (du-o-DE-num), the first portion of the small intestine.
pyloric sphincter	9-85	At the opening between the esophagus and the stomach is the _____ sphincter. At the opening between the stomach and the duodenum is the _____ sphincter.
cardiac pyloric	9-86	Chyme from the stomach passes through the pyloric sphincter into the first portion of the small intestine, which is called the _____.

duodenum	9-87	Refer to the diagram again. Chyme passes from the duodenum into the second and third portions of the small intestine, which are called, respectively, the _____ and the _____.
JEJUNUM (je-JOO-num) ILEUM (IL-e-um)	9-88	The duodenum of an average adult is somewhat less than a foot long, while the jejunum averages 9 feet in length, and the ileum averages 13 feet. Thus, the average total length of the small intestine of an adult is about _____ feet.
23	9-89	The shortest portion of the small intestine is the duodenum, which is less than one foot long. The second portion, called the _____ is about 9 feet long, and the third and longest portion, the _____, is about 13 feet long.
jejunum ileum	9-90	The large intestine, into which indigested food passes after it leaves the small intestine is approximately 5 feet long but has a greater diameter than the small intestine. In other words, the small intestine is (longer/shorter) _____ than the large intestine but has a (larger/smaller) _____ diameter.
longer smaller	9-91	Because of its great length, the small intestine is arranged in irregular coils within the abdominal cavity. The diagram is not an exact representation of the way these coils look but is a simplified drawing, designed to show the direction in which the digested food or chyme moves through the intestines. (No response required-move to frame 92)

	9-92	List the three parts of the small intestine in the order in which chyme moves through them after it leaves the stomach.
		
		        (1) _____
		        (2) _____
		        (3) _____

1. duodenum 2. jejunum 3. ileum	9-93	The small intestine contains intestinal glands which secrete intestinal juice. In the duodenum, chyme from the stomach is mixed with _____ juice produced by the small intestine.

intestinal	9-94	Chyme in the duodenum is also mixed with secretions from the two organs drawn in red in the diagram, the _____ and the _____.
liver pancreas (either order)	9-95	The pancreas, located beneath the liver and the stomach, is both an endocrine and an exocrine organ. The endocrine portion of the pancreas has been discussed in Chapter eight of this program. The pancreatic juices which enter the duodenum are secreted by the other portion of the pancreas, the (endocrine/exocrine) _____ portion.
exocrine	9-96	When chyme enters the duodenum, it stimulates the duodenum to produce a hormone called SECRETIN (SEC-re-tin). The secretin is then carried to the exocrine portion of the _____ where it stimulates the secretion of pancreatic juice.
pancreas	9-97	The pancreas, after being stimulated by the hormone _____, produces pancreatic juice.
secretin	9-98	The pancreatic juices are capable of digesting all three classes of food: carbohydrates, fats, and proteins. These juices are mixed with chyme when the chyme is in the _____, the first portion of the small intestine.
duodenum	9-99	The pancreatic juice contains an enzyme which splits complex sugars into simple sugars. This enzyme is called PANCRE-ATIC AMYLASE (AM-i-lace). The action of the salivary enzyme, _____, on carbohydrates splits complex sugars. Complex sugars are also split into simple sugars by the pancreatic enzyme called _____.
ptyalin pancreatic amylase	9-100	The pancreatic enzyme responsible for splitting fat molecules is LIPASE (LI-pace). The fact that fats are also known as LIPIDS (LIP-idz) may help you to remember that the pancreatic enzyme called _____ splits _____ molecules.

334

lipase fat (lipid)	9-101 Match the items in these two columns. (1) pancreatic amylase (a) salivary enzyme, splits complex sugars into simple sugars. (2) lipase (b) pancreatic enzyme, splits complex sugars into simple sugars. (3) ptyalin (c) pancreatic enzyme, splits fat molecules
1. (b) 2. (c) 3. (a)	9-102 Pancreatic juices are mixed with chyme while the chyme is in the duodenum. BILE, secreted by the liver, is also mixed with the chyme while it is in the _____ portion of the small intestine.
duodenum	9-103 The liver secretes a fluid called _____, which acts upon the fats in chyme.
bile	9-104 In the duodenum, the fatty elements of chyme are digested partly by the pancreatic enzyme, _____, and partly by the _____ secreted by the liver.
lipase bile	9-105 Bile, secreted by the liver, acts upon the _____ in chyme, making them easier for the body to absorb.
fats	9-106 Bile and pancreatic juice are both reservoirs of alkali. As a result, they help to neutralize the _____ in the chyme that is produced in the stomach.

acids	9-107 Thus we find that bile fills two important rolls: it acts upon (1) _____ , and in combination with (2) _____ it helps to (3) _____ the acids.
1. fats 2. pancreatic juice 3. neutralize	9-108 Bile is secreted by the liver but is stored in the GALL BLADDER, which, as you can see from the diagram, is located (above/below) _____ the liver.
below	9-109 Bile may enter the duodenum directly from the _____ which secretes it, or it may (more commonly) enter from the _____ _____ where it is stored and concentrated.
liver gall bladder	9-110 Refer to the diagram of the digestive system. Chyme, mixed with the secretions of the liver, pancreas, and small intestine is forced from the duodenum through the _____ and _____ portions of the small intestine.
jejunum ileum	9-111 The alternating contraction and relaxation of the muscles of the intestines moves chyme through them. In the intestines, as in the esophagus and stomach, the chyme is pushed along by _____ action.
peristaltic	9-112 The interior of the small intestine is covered with microscopic finger-like projections called VILLI (VIL-li). Water and the final products of digestion are absorbed into blood and lymph vessels through the projections called _____ in the _____ intestine.
villi small	9-113 The function of the villi in the small intestine is to _____ water and the final products of digestion into the blood and lymph vessels.

absorb	9-114 The carbohydrates, fats and proteins have been broken down into simple (1) _____, simple protein molecules and (2) _____ _____, and fatty (3) _____.
1. sugars 2. amino acids 3. acids	9-115 Simple sugars, simple protein molecules and amino acids, and fatty acids are now absorbed by the many small projections of the small intestine, the _____.

villi	9-116

VILLUS

CAPILLARY NETWORK

LACTEAL

BLOOD FLOW

LINING OF THE _____ _____.

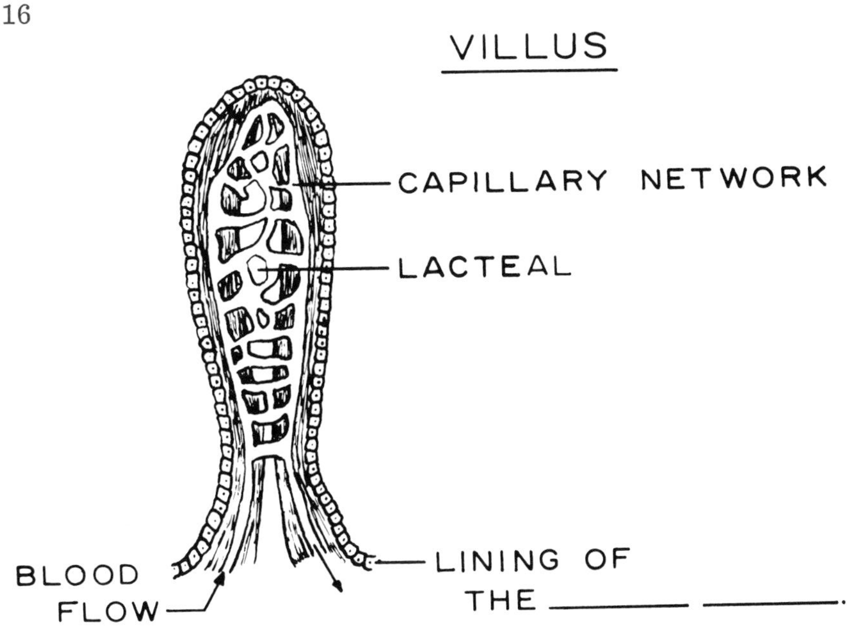

(Fill in the blanks).

small intestine	9-117 The products of digestion are absorbed by passing through the lining of the small intestine into either the _____ _____ or the _____.
capillary networks lacteals (either order)	9-118 The fatty acids are absorbed by the lacteals, and the simple protein molecules, amino acids, and simple sugars are absorbed by the _____ _____.

capillary networks	9-119　　The digestive process is completed when the food is broken down and absorbed by the villi. The simple sugars enter the _____ _____ of the villi, and the fatty acids enter the _____ of the villi.
capillary networks lacteals	9-120　　Little or no absorption takes place in the esophagus or the stomach. Absorption does not begin in earnest until food reaches the _____ _____.
small intestine	9-121　　Products of the digestive process from the small intestine then pass into the first portion of the _____ _____ or colon
large intestine	9-122　　　　　The Large Intestine 　　　　The large intestine's main function is to absorb water from the intestinal contents and then transfer the waste products with little or no _____ value from the body.
nutrient	9-123　　The first part of the large intestine is called the_____. (Refer to the diagram of the digestive system).
CECUM (SEE-come)	9-124　　Peristalsis in the large intestine, or colon, pushes the waste products up from the cecum into the (1)_____ colon, across the (2) _____ colon and down through the (3) _____ colon and the (4) _____ colon. (Refer to the diagram).
1. ascending 2. transverse 3. descending 4. sigmoid	9-125　　The first portion of the large intestine is called the _____. The second portion consists of the ascending, transverse, descending and sigmoid _____.

338

cecum colons	9-126 Waste products from the colon then pass into the last portion of the large intestine, called the _____, and are finally excreted through the _____.
rectum anus	9-127 The waste product of digestion is called FECES (FE-seez). Normally, this waste product, _____, is stored only in the sigmoid (S-shaped) portion of the _____.
feces colon	9-128 Usually feces are stored in _____ section of the _____.
sigmoid colon	9-129 Normally the rectum is empty. The desire to defecate waste occurs when the rectum becomes distended with feces. When food, water or even a cigarette stimulate the stomach, a reflex is initiated called the GASTROCOLIC reflex. The sigmoid contracts and forces feces into the rectum. This action produces the desire to _____.
defecate	9-130 Stimuli such as food or water induce the stomach to initiate a _____ reflex which in turn induces the _____ to contract.
gastrocolic sigmoid	9-131 In response to the gastrocolic reflex, peristalsis of the sigmoid colon forces feces into the last portion of the large intestine, namely, the _____.
rectum	9-132 In response to the gastrocolic reflex, feces enter the rectum and distend it. Distention of the rectum stimulates peristalsis of the entire colon which then pushes the feces out through the _____.

anus	9-133 Guarding the opening of the anus are two circular sphincters. The internal sphincter is composed of smooth muscle, and the external sphincter is composed of skeletal muscle. Thus the internal sphincter is under involuntary control, and the external one is under _____ control.
voluntary	9-134 At the opening of the anus, the internal sphincter is under (involuntary/voluntary) _____ control, while the external sphincter is under (involuntary/voluntary) _____ control.
involuntary voluntary	9-135 The internal sphincter relaxes during the reflex stimulation of the colon, thus allowing the _____ to escape.
feces	9-136 Defecation may be blocked by the voluntary contraction of the (internal/external) _____ sphincter composed of skeletal muscle.
external	9-137 In summary, it was mentioned that the main function of the large intestine or (1) _____ is to absorb (2) _____ from the intestinal contents and to then move (3) _____ products out of the body.
1. colon 2. water 3. waste	9-138 The waste products of digestion which are composed of non-nutrient materials and bacteria are called _____. One reflex which brings about the desire to remove the waste products from the body is called the _____ reflex.
feces gastrocolic	9-139 To review the entire process of digestion: The entire digestive canal is referred to as the ALIMENTARY (al-i-MEN-ter-e) CANAL. The passages and spaces through which food passes between its ingestion through the mouth and its excretion through the anus make up the _____ canal.

alimentary	9-140 Food enters the mouth, where it is ground and crushed by the teeth. The _____ pushes the food against the teeth and also propels it toward the pharynx where it can be swallowed.
tongue	9-141 In the mouth, food is mixed with _____, which moistens it and prepares it for swallowing.
saliva	9-142 Saliva contains an enzyme called (1) _____, which splits complex (2) _____ into simple (3) _____.
1. ptyalin 2. starches 3. sugar	9-143 After leaving the mouth, food passes into the _____, which is a common passageway for food and air. When food reaches the pharynx, the act of swallowing the food becomes a(n) (voluntary/involuntary) _____ act.
pharynx involuntary	9-144 Food then passes from the pharynx into the (1) _____. Closure of the glottis by the (2) _____ at this point prevents food from entering the (3) _____ or voice box.
1. esophagus 2. epiglottis 3. larynx	9-145 The bolus of food moves through the esophagus by means of a process called (1) _____ or (2) _____ action, which involves the alternating (3) _____ and (4) _____ of the esophagus.
1. peristalsis 2. peristaltic 3. contraction 4. relaxation (3 and 4 either order)	9-146 The peristaltic action of the esophagus forces the food through the (1) _____ sphincter into the (2)_____ where gastric juices mix with it to form a semi-liquid called (3) _____.

1. cardiac 2. stomach 3. chyme	9-147	In the stomach: 1. _____ _____ activates pepsinogen. 2. _____ breaks proteins into smaller molecules and amino acids. 3. _____ is secreted to protect the stomach lining.
1. hydro- chloric acid 2. pepsin 3. mucus	9-148	Chyme passes from the stomach into the (large/small) _____ intestine.
small	9-149	Chyme from the stomach passes through the _____ sphincter into the first portion of the small intestine, which is called the _____.
pyloric duodenum	9-150	After leaving the duodenum, chyme passes first through the _____, or second portion of the small intestine, and then through the _____, or third portion of the small intes- tine.
jejunum ileum	9-151	In the duodenum, chyme is mixed with intestinal juice, with _____ juice and with _____ from the liver.
pancreatic bile	9-152	Pancreatic juice contains _____, which digests fat, and pancreatic _____, which splits complex sugars into simple sugars.
lipase amylase	9-153	Bile, secreted by the (1) _____, acts upon the (2) _____ present in chyme. Bile and pancreatic juice also help to neutralize the (3) _____ in the chyme.

1. liver 2. fats 3. acid(s)	9-154 Bile is stored in the _____ _____ where it is concentrated.
gall bladder	9-155 Absorption of the nutrients and digested food takes place mainly in the (large/small) _____ intestine.
small	9-156 The interior of the small intestine is covered with microscopic projections called _____ . The function of these microscopic projections is _____ _____. (Use your own words.)
villi to absorb nutrients and water (or any synonymous statement)	9-157 The fatty acids are absorbed in the _____ of the villi. The amino acids, and simple sugars are absorbed in the _____ _____ of the villi.
lacteals capillary network	9-158 The remaining products of digestion from the small intestine pass into the large intestine. These products pass first into the _____ . The products then pass through ascending, transverse, descending and sigmoid portions of the _____ .
cecum colon	9-159 Waste products from the colon pass finally into the last portion of the large intestine, which is called the _____ , and are excreted through the _____ .
rectum anus	9-160 Under normal circumstances, feces is stored only in the _____ portion of the _____ .

sigmoid colon	9-161 One reflex which brings about the desire to defecate is called the _____ reflex.
gastrocolic	9-162 At the opening of the anus are two sphincters. The internal spincter is under (voluntary/involuntary) _____ control, and the external sphincter is under (voluntary/involuntary) _____ control.
involuntary voluntary	9-163 THE LIVER In this last section, some of the functions of the liver will be considered. The liver is essential to life. Its vital functions include secretion, the formation and storage of essential body substances, and the detoxification of blood. First, as you have already learned, the liver secretes _____, which is essential for the emulsification of fats.
bile	9-164 In addition to secreting bile, the liver stores and regulates the release of nutrients which are carried to the body cells through the _____ stream.
blood	9-165 Nutrients are absorbed through the villi of the small intestine and pass into capillaries or lacteals. The capillaries, by way of tributaries, empty into a single vein, the PORTAL VEIN, which leads to the liver. The nutrient-laden blood from the small intestine is then carried to the liver through this _____ vein.
portal	9-166 The liver stores nutrients which are mainly composed of GLYCOGEN (GLY-co-jen) and releases them into the blood stream as they are needed by the body. The nutrients processed from the digestion of food are not released immediately but are stored by the _____ until the body's output of energy requires their release.

liver	9-167 Another function of the liver, then, is to store and regulate the amount of nutrients, mainly _____ , which are released into the blood stream.
glycogen	9-168 In addition to glycogen, the liver also stores <u>HEPARIN</u> (HEP-a-rin), an anticoagulant which maintains the fluidity of the blood. This anticoagulant, called _____, is both produced and stored in the _____.
heparin liver	9-169 The liver also produces and stores <u>FIBRINOGEN</u> (fi-BRIN-o-jen) and <u>PROTHROMBIN</u> (pro-THROM-bin), which serve as coagulants by aiding in the formation of blood clots at injured body sites. Thus, the liver produces and stores both an _____ (heparin) and two _____ (prothrombin and fibrinogen).
anticoagulant coagulants	9-170 The liver also serves as a storehouse of <u>VITAMINS</u>, including vitamins A, B_{12}, and K. Vitamin B_{12}, which is stored in the _____, is essential to the formation of red blood cells.
liver	9-171 After birth and during the later stages in the development of the embryo, red blood cells are produced by the red marrow of bones. In the early stages of embryonic development, before the red marrow is able to assume this function, the liver stores vitamin B_{12}, which is essential for the production of _____ _____ cells.
red blood	9-172 The liver, by storing vitamin B_{12}, produces red blood cells in the (mature adult/developing embryo)_____.
developing embryo	9-173 Vitamin K, which is also stored in the liver, contributes to the production of prothrombin. Prothrombin, as you learned, is formed and stored in the liver and serves to assist in the _____ of blood.

coagulation (clotting)	9-174 Iron and copper are also stored in the liver. These two elements, _____ and _____ , are important in the production of hemoglobin.

iron
copper
	9-175 The liver stores (and in some cases also forms) many essential body substances and releases them into the blood stream as needed.
	
	       Select the substances stored by the liver from the list below.
	
	     a. heparin f. mucus
	     b. copper g. prothrombin
	     c. pepsin h. fibrinogen
	     d. iron i. ptyalin
	     e. vitamins j. glycogen

a, b, d, e, g, h, and j	9-176 In addition to the secretion of bile and the formation and storage of many essential body substances, the liver also serves to remove toxic, or _____ ,substances from the blood.
harmful	9-177 Many substances absorbed through the intestines into the blood stream are toxic and potentially dangerous to the body. The blood from the intestines passes through the portal vein into the liver, where it is detoxified. Thus the liver removes harmful or _____ substances from the blood.

toxic	9-178 In summary--select the functions of the liver from the list below.
	
	     (a) produces red blood cells in the embryo
	     (b) secretes bile
	     (c) regulates the storage and release of nutrients
	     (d) secretes ptyalin
	     (e) stores glycogen, vitamins, copper, and iron
	     (f) produces coagulants
	     (g) produces gastrogen
	     (h) produces anticoagulants
	     (i) detoxifies blood

a, b, c, e, f, h and i	